FROM 0 to
∞ INFINITY ∞
in
26 CENTURIES

The Extraordinary Story of Maths

CHRIS WARING

Michael O'Mara Books Limited

This paperback edition first published in 2017

First published in Great Britain in 2012 by
Michael O'Mara Books Limited
9 Lion Yard
Tremadoc Road
London SW4 7NQ

A CIP catalogue record for this book is available from the British Library.

Papers used by Michael O'Mara Books Limited are natural, recyclable products
made from wood grown in sustainable forests. The manufacturing processes
conform to the environmental regulations of the country of origin.

ISBN: 978-1-84317-873-6 in hardback print format
ISBN: 978-1-78243-767-3 in paperback print format
ISBN: 978-1-84317-921-4 in EPub format
ISBN: 978-1-84317-922-1 in Mobipocket format

1 2 3 4 5 6 7 8 9 10

Designed and typeset by www.glensaville.com
Cover design by Dan Mogford
Illustrations by Aubrey Smith
Printed and bound by CPI Group (UK) Ltd, Croydon, CR0 4YY

www.mombooks.com

Contents

Introduction

There's no hiding from mathematics. It's a subject so rich and diverse that we use it to explain everything from the Big Bang to how to improve your chances in a game show. Maths too plays an integral role in everyday life. You might work in a technical profession that demands frequent number-crunching, or you might only have to perform calculations when you're working out your accounts or comparing special offers on the Internet.

Maths is drummed into us from a young age. You have most likely received some degree of education in mathematics, probably up until the age of sixteen. You will have been taught arithmetic – how to perform calculations; geometry, which helps us to understand shape and space; and algebra, which allows us to solve problems without having to resort to trial and error.

You may be one of an increasing number of people who has studied mathematics to degree level, or beyond. In which case you may be familiar with calculus, complex numbers, mechanics, statistics, decision mathematics or any myriad mathematical field that exists.

Whatever level of mathematics you have so far reached, it is unlikely you have ever been told much of its back story. Who decided we should work in tens? Why are there 360 degrees in a circle? Who invented algebra? Every aspect of mathematics, from the numbers we use to the way modern mathematicians tackle the big unanswered problems, is the product of thousands of years of human endeavour that goes largely unmentioned in maths lessons and textbooks around the world.

From 0 to Infinity in 26 Centuries sheds light on the fascinating history of mathematics, starting with the earliest people and working forwards to the modern day. It's a chronicle of people and their cultures; their beliefs and aims. Why does the Mayan calendar end in 2012? Why were there no notable Roman mathematicians, and yet so many in Ancient Greece? When did scientists start using maths to develop theories?

I hope that you will find the stories that follow interesting, touching and entertaining. I hope too that it will help you find a new respect for mathematics and for the people that helped to develop it into the wonderful subject that it is today.

Prehistoric Maths

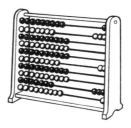

BACK TO THE BEGINNING

It has been estimated that the earliest humans arose in Africa approximately 250,000 years ago. These people left behind little evidence of their existence other than a few fossils, so we know very little about their culture, if indeed they had one.

So, what can we say about their mathematical abilities?

Early estimates

A trait that all humans – and indeed primates and some other animals – have is the ability to **subitize**: to know at a glance how much a small number of things amounts to. Here is an example:

III

If you have the ability to subitize, you will be able to look quickly at the lines above and spot that there are three of them, without having to count each line. Now try this one:

llllllllllllllllllllll

There are twenty-three lines here, but I only know that because I typed them. At a glance, the best you would probably be able to do is to say that there are 'around twenty' or 'two dozen' lines. The instantly recognizable and countable pips on a die are a modern-day example of subitizing.

We think that subitizing is a trait that has evolved in animals to allow them to make quick decisions with regard to fight-or-flight-type situations: one or two wild dogs and you might be happy to stand your ground (as long as there's a stick nearby that you can use to fend them off); three or more dogs and you're likely to run to the nearest tree.

You and I are literate and numerate humans who can't remember what it was like not to be able to count. We see three lines and we cannot help but think of the number three. Our first ancestors, however, would have had no word for three and, perhaps more significantly, possibly no concept of three as a number.

THE STONE AGE

Approximately 200,000 (notice the comma to help you subitize all those zeros!) years after they first walked this earth, humans gained what anthropologists call 'behavioural modernity': they started doing things that differentiated them from other animals. They developed language, tools, cooking, make-believe, painting, and had begun to ponder the nature of existence and all the other things that make us human. These were the Stone Age hunter-gatherers. We know a touch more about their mathematics because the remains of cavemen types were unearthed from the nineteenth century onwards and written about by their pith-helmeted discoverers.

A counting controversy

The Ishango bone is the thighbone of a baboon that was discovered in the Democratic Republic of the Congo, Africa, in 1960. Dated at approximately 20,000 years old, the bone has caused much controversy among scientists. The bone has three sets of grooves carved deliberately into it, and if you count the grooves you find the following sequences: (9, 19, 21, 11), (19, 17, 13, 11) and (7, 5, 5, 10, 8, 4, 6, 3). Some scientists believe that this is evidence not only of the Stone Age peoples' ability to count up to numbers much higher than the more recent Aboriginal tribes (well, higher than three anyway; see box on page 12), but that the numbers in each set shows evidence of an understanding of counting in tens, odd numbers and prime numbers. This argument has been challenged by other scientists who suggest the grooves were either decorative or intended to make the smooth bone easier to grip, and therefore mathematically meaningless.

Modern-Day Hunter-Gatherer Tribes

The Pirahã tribe lives today in the Amazon rainforest. They are consummate experts at jungle survival. The tribe's language is so simple that its hunters use a whistled version of it while out trailing game. Remarkably (at least to us), their language contains no numbers and, despite trading commodities such as T-shirts, metal knives and alcohol with other tribes and river traders, the Pirahã show no inclination to adopt a number system either. These people live in such a way that numbers have no function for them – they live hand to mouth in the equatorial rainforest, where food is available all year round.

Australian Aboriginal tribes were living in a hunter-gatherer society when they were first encountered during the eighteenth century. The tribes that possessed a concept of numbers generally had words for one, two and sometimes three. Any numbers larger than three they made by adding together a combination of the first three numbers. So a tribe with words for one, two and three would have been able to count to nine by saying: one, two, three, three-one, three-two, three-three, three-three-one, three-three-two, three-three-three. The fact that these people had no word for numbers larger than three suggests that they very rarely, if ever, needed to use them.

The Ishango bone notwithstanding, it is fair to say that many Stone Age tribes would have had a fairly childlike grasp of numbers. And, like children today, we can be fairly sure that our early ancestors used their fingers for counting – a key development along the way to numeracy.

COUNTING ON FINGERS (AND OTHER BITS TOO!)

We humans have eight fingers and two thumbs, and if you watch any young child learning to count or add (counting is in fact just adding on one each time), you'll see that these convenient ten counters are too tempting not to use. Consequently, we instinctively like the idea of numbers coming to us in batches of ten.

A means of communication

We can also use our fingers to communicate numbers non-verbally, which is as useful now when you are in a foreign country as it was for Stone Age hunter-gatherers. They perhaps would have used their fingers to express numbers that they didn't have words for, or to communicate an idea to other people across a language barrier.

The upper limit for counting on your fingers is ten, which, as we have seen, would have been more than enough for many Stone Age people. As societies developed, larger numbers were required, yet counting in batches of ten continued. The modern-day words we use for numbers usually have their roots in this tradition – the English words 'twenty' and 'thirty' come from 'two-tens' and 'three-tens' respectively. Some ancient cultures such as the Mayans and the Celts used fingers and toes to count in batches of twenty rather than ten, the evidence of which still exists in some languages today. The French word for eighty, *quatre-vingts*, literally translates as 'four-twenties'; the Welsh language uses a similar system: thirty-one in Welsh is *un ar ddeg ar hugain*, which is 'one on ten on twenty'.

When you count on your fingers, or, more importantly, use them to show that Greek barman how many shots of Metaxa you'd like to buy, it does not really matter which fingers you use – we tend to use what is anatomically more comfortable. For example, it is far easier to show the number four using the fingers of one hand rather than a thumb and three fingers on the same hand (try it!). It does mean, however, that there are ten different ways of showing one on your fingers, and that the finger system relies on a person's ability to count up all the fingers that are shown.

Some cultures have navigated these potential pitfalls by assigning a value to different parts of their body. On the Torres Strait Islands between Australia and Papua New Guinea, thirty-three different body parts are used for counting, and indeed as words to represent the numbers. For example, to a Torres Strait Islander the ring finger on your left hand means sixteen, your right shoulder means eight, your left knee twenty-four and your little toe on the right foot thirty-three. This system has evolved to allow for effective communication between islands that are home to several different languages.

The Legacy Lives On

With our highly evolved modern-day number systems, we have little need to communicate numbers with our hands, but we do still see it occasionally. Traders on the Stock Exchange floor have the 'Open Outcry' communication system, which involves shouting and hand signals, to buy and sell shares in a noisy trading pit.

The Stone Age humans persisted in their ways, hunting and gathering without the need for many numbers at all. However, approximately 10,000 years ago, people in fertile areas around the world's great rivers decided to settle down and get civilized. This led to the need for much bigger numbers, methods of recording them and every schoolkid's favourite – arithmetic.

Early Civilized Maths

FROM HUNTER-GATHERERS TO HERDERS

According to historians there are five main stages involved in a society becoming 'civilized'. The first stage is the ability to make and control fire – *Homo sapiens* and their ancestors have been creating fires for approximately half a million years. The second stage is the cultivation of crops, which really requires the help of domesticated animals. The Neolithic Revolution, which occurred in independent locations across the globe approximately 10,000 years ago, saw humans begin to stay in one place, grow crops and domesticate and rear livestock.

Counting sheep

The first shepherds would have needed a method for counting their animals, so it seems obvious that they would have been good at counting. Or would they? Equally, the first farmers would have needed a method of working out what time of year it was so they knew when to plant their seeds – surely more numbers were involved here?

The Neolithics used a system called pebble counting. In order to count their herds of sheep, the shepherds placed one counter – perhaps a pebble or a fruit pip – in a bag for each member of the flock. To find out if all of the animals in the herd were safe the shepherds then removed a pip for each sheep counted. If by the time they got through all of the sheep they still had pips remaining, they knew then that one or more of the sheep had been lost. Hopefully, if they were half-decent shepherds, this number would have been one of the low numbers (1–10), which we now know hunter-gatherers were equipped to deal with. The farmers would have used a similar system to count from a key event, such as the rains beginning, or the birds flying south, to let them know when to plant their crops.

The economy takes shape

Surplus is an inevitable consequence of agriculture and the domestication of animals. Once there was an excess of food, early civilizations could start to trade their surplus, which in turn promoted the idea of value. This is the third sign of civilization – the idea of an economy; of goods being bartered or sold. Naturally, they needed a way to cart around their surplus of goods, so the fourth sign of civilization is the wheel, which seemed to be widespread by *c*. 4000 BC.

THE DAWN OF THE ACCOUNTANTS

With a reliable surplus of food it then became possible for some people to fill their time doing things other than finding or growing food in order to survive. By *c*. 3000 BC towns and cities were filled with such people. These large urban societies needed organizing in order to function properly, and the ability to count large numbers was key to this. The need also arose for these new, large numbers to be recorded – and hence written – for the first time. Among the myriad flagship new professions that arose – craftsmen, soldiers, farmers, merchants – a new literate class – the scribe – was to be found; almost certainly many of these scribes were numerate accountants and – inevitably – tax collectors.

The Bronze Age

The fifth sign of a civilization is the use and working of metals, which started at approximately the same time as the first towns and cities arose. The easiest metal to work is copper, which is used to make bronze. Hence we call this period in history the Bronze Age.

Most of the earliest civilizations developed around the fertile areas near rivers, where the land was suitable for farming and raising livestock. Three notable civilizations that arose during this period, and about which we know a great deal, were:

1. The Mesopotamians: a collective name for the Sumerians, Akkadians, Babylonians and Assyrians. They lived in the Middle East from *c.* 3000 BC.

2. The Ancient Egyptians: the baddies in the Old Testament. They were based along the River Nile from *c.* 3000 BC.

3. The Mayans: from Central America. Their earliest stages of development began *c.* 1500 BC; they entered a stage of development similar to the Mesopotamians and Egyptians after AD 250.

As some cities rose in importance and began to dominate the surrounding area, either economically or by strength of arms, certain regions began to adopt the most effective numbers system. Unfortunately the early records from many such cultures have not survived, perhaps because these people were sited next to large rivers that flooded often. But if we take a quick look at some of the highlights of the archaeological findings that *have* survived from each civilization, we gain some idea of an evolution of written numbers and number systems from our earliest urbanite ancestors.

MESOPOTAMIAN MATHEMATICS

Mesopotamia means 'between rivers' in ancient Greek, and refers to the cultures that sprang up between the Tigris and Euphrates rivers in present-day Iraq – a very fertile area of land often called the Cradle of Civilization. Despite their early beginnings, we know a fair amount about the Mesopotamians because they performed all of their writing on clay tablets, which are hardy enough to withstand repeated soakings.

Clay tablets, however, are not easy to write on. The Mesopotamians began with a pictographic language, when the written symbol for a word is a stylized picture of the thing being described. However, drawing decent images in thick wet clay is tricky, so they took to using the end of a wedge-shaped stylus (a rod-like implement with a pointed end) to make marks.

Number system

The Mesopotamian's number system was base 60 (also referred to as **sexagesimal**), which means they counted in blocks of 60 rather than in blocks of 10, as we do today. Lots of numbers go into 60 (mathematicians would say 60 has many factors), which makes it a convenient number with which to do arithmetic. We still see a few reminders of it today – 60 seconds in a minute and 60 minutes in an hour hark back to the Babylonians.

Cunningly, their number system contained only one symbol, made using the end of a stylus:

They would use up to nine of these symbols. They would show a 10 by rotating the stylus by 90 degrees to get a slightly different symbol:

So the number 47 would look like this:

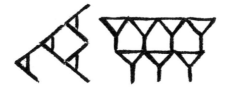

The Mesopotamians could write up to the number 59 in this way, using a sub-column for each of the tens and units digits. To write numbers larger than 59, they would write a new number alongside the tens and units (just as we can go as high as 9 in one column before we have to move on to the next one). Whereas our columns follow the pattern units, tens, hundreds, thousands, etc., according to our way of thinking the Mesopotamians' columns went units, sixties, three-thousand six-hundreds, two-hundred and sixteen-thousands. A Mesopotamian would think of our number 437 as being 7 lots of 60 (7 × 60 = 420) plus 17, which they would write like this:

Zero interest

Although this seems like a pretty decent number system, not unlike our own, there were a couple of problematic areas. The first problem was that until *c.* 500 BC the Mesopotamians had no symbol for zero, which means they had no way of showing an empty column. For example, if I write down 205, the zero tells you that I mean two-hundred and five, and nothing else. The Mesopotamian number system was flawed because empty columns in the middle or at the end of a number were missing. For example:

This looks like 60 + 10 = 70. But there could be an empty column in the middle, in which case the number would be 3,600 + 0 + 10 = 3,610. Or there could be an empty column at the end, in which case we would have 216,000 + 60 + 0 = 216,060 – quite a large difference. Apparently, Mesopotamians tended to rely on the context in which the numbers were used in order to read them in the most reasonable way.

Early Arithmetic

Many people consider the abacus to have been the ancient world's version of the electronic calculator. In fact, the counting frame with beads – which most people think of when they hear the word abacus – is a relatively modern piece of technology, first made popular in China after AD 1000.

The word 'abacus' is thought to come from the Hebrew word for 'dust', and the first abacuses were simply that – a board or level surface strewn with dust that could be used as a scratch pad for calculations. Eventually the dust was replaced by a board with tokens that could be placed in columns to allow for the addition of large numbers without having to be able to count higher than ten.

Later, the Romans used pebbles or, in Latin, *calculi* (from which we get the words calculus and calculation). In England, we called the tokens 'counters', which is why shops had a countertop to put their counting board on.

Multiplication madness

The second problem arose when the Mesopotamians tried to multiply numbers together. Whichever way you multiply using our decimal system, you need to have memorized your times tables up to 9 x 9 (because 9 is the highest digit we have). However, according to the Mesopotamian system, you needed to know your times tables up to 59 x 59! We think they used a few key times tables, written on small tablets, to help, but even so their times-table tests at school must have been a nightmare.

Archaeologists have found many hundreds of clay tablets littered with Mesopotamian mathematics. It seems the Mesopotamians were able to use fractions, to work out the areas of rectangles and triangles, and to solve quite complicated equations. My favourite fact is that many tablets found appear to have been maths homework! But maybe you need to be a maths teacher to appreciate that ...

ANCIENT EGYPTIAN MATHEMATICS

The ancient Egyptians were a talented lot. In addition to building the pyramids, many of which are still standing over 4,000 years later, they also turned their hands to committing the written word to papyrus, a paper-like material made from interwoven reeds. Papyrus was a much more forgiving material to write on than the clay tablets the Mesopotamians were using further north. As such, unlike the Mesopotamians, the ancient Egyptians were not limited to using a single symbol. However, because papyrus rots, especially if it gets wet, it does mean that the vast majority of the writing of ancient Egypt has been destroyed over time.

Systems in place

It also seems that the Egyptians were not limited to one writing system.

Egypt is famous for its hieroglyphics – pictograms they carved on to their monuments, and which remained a complete mystery until French soldiers unearthed the Rosetta Stone in 1799.

Hieroglyphics were the ancient Egyptian equivalent of calligraphy – decorative writing for use only on wedding invitations and inscriptions. The Egyptians had another writing system called hieratic, which they used for everyday stuff – a much easier and faster way to write script that scribes would then use for their calculations.

Hieroglyphic numbers had symbols for 1, 10, 100, 1,000, etc., which the ancient Egyptians would combine to make the required number:

1   10   100 1000

So if an Egyptian wanted to refer to Rameses' 1,234 chariots on his latest obelisk, he would have used the following symbols:

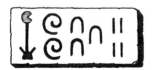

Because they added up the symbols in order to generate a total, the Egyptians could write the symbols in any order and direction they pleased – a handy tool when they wanted to be decorative.

The hieratic number system was a little more complicated because it used different symbols for each unit, each ten, each hundred and each thousand. The symbols for 40 and 50, for example, bore no relation to each other. It seems that this system relied on the fact that the scribes would be familiar with the symbols and would be able to perform calculations either in their heads, or by converting to the hieroglyphic system for tricky sums.

The business of numbers

Three important sources of mathematical information were left behind from ancient Egypt: the Rhind Papyrus, the Moscow Mathematical Papyrus and the Berlin Papyrus. All three documents contained mathematical problems in arithmetic and geometry, alongside, interestingly, the first written information about pregnancy tests.

From these three sources we have learned that the Egyptians used fractions. However, they only used fractions that had a numerator of 1 – that is, the number on the top of the fraction could only be 1. They would talk about more complicated fractions by adding these **unit fractions** together. So, for example, they would think of ¾ as ½ + ¼. Although slightly cumbersome, this method stood the test of time – unit fractions were still used by mathematicians in medieval times.

The pyramid builders obviously had a pretty good grasp of

geometry; the papyri contain detail about how they set about making these ancient structures. The pyramids were made with stacks of stone blocks in layers, and the steepness of a pyramid depended on the size of the overlap between two layers – the larger the overlap, the steeper the pyramid. The Egyptians devised a series of methods to work out what size of overlap was needed for different gradients. It has also been suggested the Egyptians had some idea of Pythagoras' theorem (see page 39), which would have enabled them to work out the third length of a right-angled triangle if they knew the length of the other two sides.

There are, of course, many more far-fetched theories regarding the ancient Egyptians, including the super-high technology they appropriated from the legendary island of Atlantis (or from aliens, or time travellers...). I cannot say whether such things were true, but I do know the Egyptians were pretty clever fellows.

A Tall Order

The fact about the ancient Egyptians that I always find most extraordinary is that the Great Pyramid, which was completed c. 2560 BC, was the tallest building in the world until the central towers of Lincoln Cathedral were raised in AD 1311 – that's the best part of 4,000 years!

THE MAYANS

By the first millennium AD Mayan civilization had reached a level of cultural and mathematical development similar to the Mesopotamians and the Egyptians. They declined somewhat as time went by, but when the Spanish conquistadores arrived in the early 1500s the Mayans had managed to recapture their previous levels of sophistication.

Born in isolation

The Mayans left behind a raft of evidence that demonstrated how they conducted their mathematics, but unfortunately virtually all of it was destroyed when the Spanish invaders arrived and sought to convert the region's heathens to Catholicism. *The Dresden Codex* is one of three surviving examples of Mayan writing. Although it was badly damaged during the Second World War, the book still contains a great deal of insight into the Mayan development of mathematics. Many surviving monuments in modern-day Mexico and Guatemala contain numerical information, such as dates, inscribed upon them.

Unlike the cross-pollination that occurred between the Mesopotamian and Egyptian cultures, the Mayans developed in complete isolation. They also failed to fulfil the last two criteria of 'civilization': they did not possess the wheel, perhaps because there were no beasts of burden in the parts of Central America where they lived; they also did not seem to be able to smelt metal. However, despite technically still existing in the Stone Age, the Mayans were able to build great cities, some of which contained populations of over 50,000 people.

Number crunching

So, what of their mathematics? *The Dresden Codex* is concerned only with astrology and astronomy, so everything we know about the Mayans' mathematics is shone through this lens.

The Mayans used a base-20 system, within which lay a base-5 system (much like the Mesopotamians' base-60 and base-10 system).

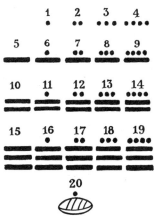

Like modern mathematics, Mayan mathematics had a grasp of **place value**: the value attached to the position of each digit. Unlike modern mathematics, Mayan mathematics placed numbers in vertical stacks, with the highest place value positioned at the top. Because the Mayans counted in groups of twenties, each level in the stack was twenty times the value of the level below. From the bottom it went something like this: 1s, 20s, 400s, 8,000s, etc. So our number 8,577 is one 8,000, one 400, eight 20s and seventeen 1s, which in Mayan looked like this:

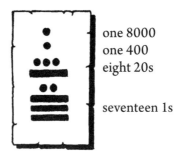

one 8000
one 400
eight 20s

seventeen 1s

The Mayans had a symbol for an empty level in the stack, e.g. they possessed a concept of zero, which avoided the confusion faced by the Mesopotamians. So the number 419 (one 400, zero 20s and nineteen 1s) looked like this:

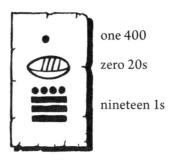

one 400

zero 20s

nineteen 1s

The Mayan calendar

As the information contained in *The Dresden Codex* would attest, the Mayans' way of life was governed by astrology. Ritual human sacrifice was an integral part of Mayan culture and was thought to aid the continuation of the Mayan people's cosmology.

Tasked with working out which rites were necessary to appease the gods, high-ranking priests were responsible for interpreting the positions of the sun, the moon and Venus. They developed a system of several different calendars, which the Mayans used in parallel. They possessed a 365-days-a-year civil calendar called the *Haab*, which comprised 18 months of 20 days each, plus 5 'nameless' unlucky days, called *Uayeb*, to make up the full total.

The Mayans' main calendar was called the *Tzolk'in*, which worked on a 260-day cycle. The Mayans devised a 13-day week and believed that 20 gods were each associated with a day of the year, and so $13 \times 20 = 260$ days in a cycle. The *Tzolk'in* was their everyday calendar, which they used to keep the date.

The 260-day cycle and the 365-day year would start together every 18,980 days, after 73,260-day cycles or 52 'vague' years, as they were called. Fifty-two years was considered to be a good, long life in those days, so in order to record anything longer than this the Mayans used yet another calendar – the *Long Count*. This calendar was used for recording dates of important events, such as kings dying or volcanoes erupting; these were the dates they chiselled on to temples and statues using their stone tools. Considering their 360-day year (ignoring the *Uayeb*, the 5 nameless days – they did not want to bring bad luck to their monuments!) the Mayans, as base-20 people, deduced that 20 of those years made something called a *k'atun*, and 20 *k'atuns* comprised a *b'ak'tun*. A *b'ak'tun* was approximately 395 years. The Mayans needed a starting point for their dates – much as we use the birth of Christ for ours – which they decreed was 3114 BC. All important dates were measured forward from that point.

The End of the World

It just so happened that the last *b'ak'tun* finished in December 2012. Some people believed the Mayans, in their infinite wisdom, predicted the world would end on this date. However, these people didn't realize the Mayans did in fact have a few more dates up their sleeves (if they had sleeves, that is) and that their calendar could be extended up to 367 million years. So they need not have worried about the world ending just yet.

The Mesopotamian, Egyptian and Mayan cultures had many things in common. Mathematically speaking, their work with numbers was functional, a means to an end – whether that end was taxation, working out when the next eclipse was due or how to build a pyramid. Maths was certainly never performed for its own sake. The Mesopotamians and the Egyptians did amass a large body of knowledge, which our next civilization – the ancient Greeks – built upon.

The Gregorian Calendar

Since ancient times the idea that a year comprises 365 ¼ days has been well accepted. This system worked well for the Romans after Julius Caesar instigated the Julian calendar in 45 BC. They even had a leap day every four years too.

But things started to get a little off kilter when it was observed that fixed points in the year, such as equinoxes and solstices, did not occur on the same day each year as time progressed. The reason for this is that a year is actually eleven minutes shy of that quarter day – not a great deal of difference, but over hundreds of years it built up to become quite an error.

By the sixteenth century the error had totalled ten days, which Pope Gregory XIII wouldn't stand for. The majority of the Catholic countries in Europe changed to the new Gregorian calendar, which got things back on track. Britain, as ever, mistrusted this newfangled European enterprise and stuck with the old calendar until 1752, by which point we had to jump from 2 September to 14 September. The Russians kept the old calendar until the communist October Revolution in 1918, which, according to the new calendar, actually happened in November.

The Ancient Greeks

$$a^2 + b^2 = c^2$$
$$c = \sqrt{a^2 + b^2}$$

Now it's time to move on to the Classical period, when the great empires of the Greeks and, later, the Romans dominated vast swathes of the known world. Their respective legacies were huge and their ways of doing pretty much anything were adopted and used for many hundreds of years after their demise.

Because we have only recently been able to translate and understand the cultures of the Mesopotamians and the ancient Egyptians, it was for a long time believed the Greeks had been responsible for the ancient world's greatest discoveries and inventions.

THE RISE OF THE PHILOSOPHERS

The philosophies of Socrates (*c.* 470–399 BC), Plato (427–347 BC) and Aristotle (384–322 BC), whose influence as mathematicians is explored later in this chapter (see page 44), were so significant that their modes of thought, minus their pagan beliefs, were later used by Christian theologists to expound their doctrine. So it is no wonder that these three philosophers of ancient Greece were held in awe for so long in Europe – their ideas, cobbled together with stories from the Bible, were held to be the literal truth, and to disagree with them publicly was unwise.

However, despite their influence, I think they developed some quite strange ideas.

Illogical logic

The first philosophers (which means 'lovers of wisdom' in Greek) were often generalists because at that time they did not possess the specializations of science and the humanities that we do today. Some of these philosophers used logic in its purest sense against clear evidence to the contrary. Zeno of Elea (*c.* 490–430 BC) developed a series of paradoxes to help explain that motion was impossible. He argued, logically, that the great war hero Achilles could never catch up with a tortoise because, having started the race 100 metres ahead, the tortoise would always be making slow progress as Achilles tried in vain to catch up. Zeno also suggested that an arrow fired from a bow was stationary because it could not be in two places at once. During its flight the arrow is constantly occupying a whole bit of space, and is therefore, in that instant, motionless. This *reductio ad absurdum* method was given credence because it 'proved' that we should not trust the evidence of our senses, which were imperfect, whereas reasoning and logic were considered to be flawless. Hmmm.

Greek mathematics

Because the Greeks were so interested in pure logic, they had a keen interest in maths for its own sake.

The ancient Greeks split mathematics into two camps: **arithmetic** and **logistics**. Arithmetic, what we today call **pure mathematics** – the study of abstract ideas rather than simple sums – was the sole preserve of intellectuals, the equivalent of today's post-graduates. However, logistics, performing calculations, was an inferior trade that was better left to numerate slaves.

The Greeks used two number systems. The first, in use from *c*. 500 BC, was the forerunner to the Roman system (see page 59), only it used Greek letters rather than Latin: I for 1, Π for 5, Δ for 10, and so on.

The second system, which replaced the first by *c*. 100 BC, was still based on the letters of the alphabet. The first ten letters, alpha (α) to iota (ι) represented numbers 1 to 10. After this point the letters went up in tens, so the eleventh letter, kappa (κ), stood for 20, and so on until rho (ρ), which stood for 100. The remainder of the alphabet then went up in hundreds. So the number 758 looked like ψνη. This number system still didn't allow for calculations to be performed with the numbers themselves, so we believe that sums would still have been carried out using counters. Despite the limitations of these numbers, they were used in Europe for over 1,000 years.

The Greek Alphabet

The Greeks, with the body of knowledge they acquired from their forebears in Egypt, Mesopotamia and elsewhere, wrote down and formalized many mathematical concepts that have been in use ever since, and which we will now explore.

Thales
(c. 624–c. 545 bc)

One of the first Greek philosophers, Thales (pronounced Thay-leez) hailed from present-day Turkey. Often considered to have been the first true scientist, at some point around 600 BC Thales began to try to explain what he saw around him in terms of natural phenomena, rather than through the actions of deities.

When it came to mathematics, Thales was, like many ancient Greeks, interested in geometry. He understood the principle of **similar triangles** and used it to predict the height of the pyramids.

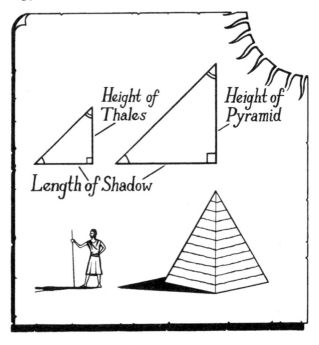

The two triangles on the preceding page are similar triangles because their angles are the same, and therefore their sides must be in proportion to each other. Thales was able to determine the height of a pyramid by measuring the length of its shadow. He waited until his own shadow was the same length as his height to measure the pyramid's shadow in order to determine how tall it was.

PYTHAGORAS (c. 570–c. 490 BC)

Pythagoras' name is instantly recognizable because his theorem has been taught to students of mathematics the world over. Because he left no written works behind, everything we know about Pythagoras was written long after he died. He founded a religious movement called Pythagoreanism, hence much that was written about him, embellished over time, assumed a decidedly mystical tinge. Among other things, Pythagoras is described as having a thigh made of gold and the ability to be in two places at once.

The theorem

Today Pythagoras is best known for his theorem for working out the hypotenuse (the longest side, the side opposite the right angle) of a right-angled triangle. In words, **Pythagoras' theorem** is described as:

> The square on the hypotenuse is equal to the sum
> of the squares on the other two sides.

But it's much pithier to say:

$$h^2 = a^2 + b^2$$

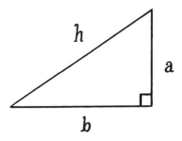

Pythagoras wasn't the first to explore the theorem. The ancient Egyptians had investigated this idea, and the Mesopotamians too – some of their homework tablets were Pythagoras-type questions (see pages 27 and 24). We know from several different sources that, as a young man, Pythagoras travelled extensively around the Mediterranean, and possibly even further afield, to gather knowledge, so it seems he had plenty of opportunity to glean this information from elsewhere.

So why is the theorem named after him? Because Pythagoras was the first Greek identified with the concept. And, since we were unable to read hieroglyphics or cuneiform until recently, it was assumed Pythagoras must have worked the theorem out independently.

Many Strings to His Bow

Another legend associated with Pythagoras is that he was the first person to work out the relationship between a length of string and the note it produces when plucked. He also noted that if the lengths of two strings were in a whole number ratio to each other then a harmonious chord was produced.

1:1
'Unison'

2:1
'Octave'

3:2
'Fifth'

4:3
'Fourth'

Pythagoras is also often attributed to the discovery of the **Platonic solids** and the **golden mean**, both of which were recorded by Plato (see pages 45–46).

Pythagoreanism

Pythagoras' religious cult was a somewhat bizarre group. Its members favoured an ascetic lifestyle by avoiding talking and following a vegetarian diet. They were a highly secretive group – the revelation of a cult secret was punishable by death. Pythagoreans were also very exclusive, and managed to antagonize the inhabitants of nearby towns enough for the populace to burn down their meeting place, killing many of the cult's members in the process.

Central to the Pythagoreans' doctrine was the idea that numbers were divine. They also believed that all numbers could be written as fractions. One notable legend centres on an unfortunate fellow called Hippasus, a Pythagorean who was pretty certain he had come across numbers that could not be written as fractions. In some of the more fanciful legends, Pythagoras asks Hippasus to take a boat out to sea with him to discuss his heretical ideas – only Pythagoras comes back. An alternative version sees Hippasus being drowned by the gods for his crimes against holy numbers.

An irrational discovery

Whatever the circumstances of Hippasus' sticky end, he may have been one of the first people to discover **irrational numbers** – numbers that cannot be written as a fraction, and whose decimal equivalent goes on for ever without repeating.

See if you can follow some good ol' Greek *reductio ad absurdum* (see page 36) to see why $\sqrt{2}$ must be an irrational number.

$\sqrt{2}$ is the square root of 2 – the number that when multiplied by itself gives an answer of 2.

$\sqrt{2} = 1.4142135623...$

If $\sqrt{2}$ can be written as a fraction, let's say it is x/y in its lowest terms.

If x/y is in its lowest term x and y cannot both be even because if they were even you would be able to divide both x and y by 2, so they could not have been in their lowest terms to start with.

If you square everything you get $2 = x^2/y^2$. This means that x^2 must be twice y^2 and so x^2 must be even because it is two times something. This in turn means x must be even because odd x odd = odd.

If x is even then y must be odd because, as you might recall, x/y was in its lowest terms and the two, therefore, cannot both be even.

If x is even, then it must be divisible by 2. So let's say x = 2 × w.

If x = 2 × w and x^2 must be twice y^2, then $4w^2 = 2y^2$, so $2w^2 = y^2$, and so y^2 must be even because it is twice something. It follows that y must be even, which conflicts with our earlier deduction that y must be odd!

If x is even, then y must also be even. But we said it must be odd. So $\sqrt{2} \neq x / y$ so $\sqrt{2}$ cannot be written as a fraction.

Socrates (c. 470–399 bc), Plato (427–347 bc) and Aristotle (384–322 bc)

Three of ancient Greece's most renowned philosophers, Socrates, Plato and Aristotle are often mentioned together because Socrates taught Plato, who in turn taught Aristotle. They were hugely influential to Western thought because, essentially, they were responsible for inventing it.

Socrates

While he did not contribute to mathematics directly, Socrates did supply a way of thinking about problems, called the **Socratic method**, which provided a logical framework for solving mathematical conundrums. Using the Socratic method, a difficult problem could be broken down into a series of smaller, more manageable pieces; by working through these smaller challenges the inquirer would eventually reach a solution to the main problem. Although Socrates generally used this method to solve ethical questions, it is equally useful for mathematical and scientific problems.

Plato's dialogues

Plato was a student of Socrates', and is well known for writing a series of works called the *Socratic Dialogues*, which use a fictional discussion between Socrates and a range of other people to set forth ideas and philosophies; a little like reading a fictional transcript of a lesson, with a student questioning the ideas put forth by his teacher.

In one such dialogue, *Timaeus*, written in *c.* 360 BC, Plato discusses several important mathematic and scientific ideas.

Plato's solids

The elements is the first topic addressed in *Timaeus*. Today modern atomic theory tells us that there are over 100 elements that can be combined to create all known substances. In his dialogue Plato was the first to propose that the four elements – fire, air, water and earth – each assume a specific shape. We name the shapes of these elements the **Platonic solids** in his honour.

The Platonic solids are 3D shapes (polyhedrons) whose faces are made up of regular (all sides and angles are equal) 2D shapes (polygons). For example, a triangle-based pyramid made up of equilateral triangles is a Platonic solid called a tetrahedron.

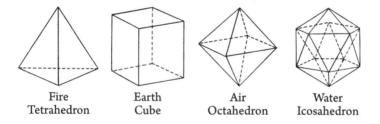

| Fire | Earth | Air | Water |
| Tetrahedron | Cube | Octahedron | Icosahedron |

These four elements, much like the elements we know today, could be combined to make any substance. There is one other

Platonic solid – the twelve-sided dodecahedron– which was not an element, but which represented the shape of the universe.

Dodecahedron

Going for gold

Another important concept discussed in *Timaeus* is the **golden mean**, sometimes called the golden ratio or golden section.

The golden mean is the optimum position between two extremes, and it's also a number: 1.6180339887... – one of those irrational numbers the Pythagoreans were not very keen on. Much like √2, the golden mean cannot be written as a fraction because its decimal continues for ever without repeating. This is inconvenient when it comes to writing the number down, so mathematicians use the symbol φ (the Greek letter phi) to represent it.

Another irrational number that has its own Greek letter is 3.14159265...: π, which you will remember from learning about circles at school. We get π by dividing a circle's circumference (its perimeter) by its diameter (the distance across the circle through the centre). It doesn't matter what the size of the circle is, you always get the same value: π. φ has a similar geometrical provenance.

If a line is divided into a longer part and a shorter part, and if the total length $(x + y)$ divided by x gives the same value as $x \div y$ then the line has been split into the golden ratio. As with π, the length of the line doesn't matter – to get it to work you find that $x \div y = 1.618... = \varphi$.

The same idea works with shapes too. The ancient Greeks considered a rectangle with its longer side φ times longer than its shorter side to be the most aesthetically pleasing rectangle possible.

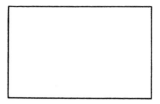

It is often said that many important examples of sculpture and architecture are made using the golden mean.

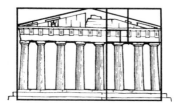

The Parthenon was designed according to the golden mean. Its length and height and the space between the columns were designed in perfect proportion to one another.

Aristotle

The son of the doctor at the court of the kings of Macedon, Aristotle was a nobleman who became a hugely influential philosopher. He was taught by Plato and later became a teacher at Plato's Academy, and he contributed ideas on a whole host of subjects, from politics and ethics to physics and zoology. So wide-ranging were his skills, it has been suggested Aristotle knew *everything* it was possible to know. Indeed, his influence extended through to the philosophy of the modern world.

Zero Option

Aristotle let things slip in his treatment of numbers. He felt that a number only really had meaning if it was an amount of something: a pile. In Aristotle's eyes, 10 apples, 1 apple, ½ an apple and ⅒ of an apple were all valid numbers. However, if you do not have an apple, you have nothing to pile up or count – zero, as far as Aristotle was concerned, was not a number.

Aristotle is known chiefly for his logic, a series of works that comprised the earliest-known study of the theory of logic. His theories have since split into many different branches, some highly mathematical, others more philosophical.

Aristotle's work in mathematics and science focused on explaining the way things behave by describing them rather than using numbers and equations. He was among the first to explain the motion of objects (a subject we today call kinetics, from the Greek for 'movement'). Aristotle's descriptions acknowledged that time and space are not arranged in indivisible chunks but are continuous, which allowed him to show that Zeno of Elea's ideas were flawed and that Achilles would have been able to catch up with the tortoise!

EUCLID (c. 325–c. 265 BC)

While little is known about the Greek mathematician Euclid, we do know that he was active in Alexandria in Egypt under Greek rule, and is notable for having penned a groundbreaking book called *Elements*. Certainly one of the most important maths books of all time, Euclid's *Elements* was considered essential reading for any scholar well into the nineteenth century.

Elementary proof

Although Euclid drew on the ideas of others, he was one of the first mathematicians to produce work that used mathematical logic in order to prove theories. This idea of proof is one of the foundations of mathematics.

Elements covers much of geometry and ideas about numbers, including prime numbers and other number sequences, and all of Euclid's geometrical constructions were made using only a pair of compasses and a straight edge.

It is split into thirteen books, each of which starts with

definitions of words to help make it clear what Euclid means when he refers to words such as point, line, straight, surface, etc. Euclid then sets out a list of axioms or statements that are evidently true, such as 'all right angles are equal to each other' and 'if A=B and A=C, then B=C'.

The next section of *Elements* is called 'Propositions', in which Euclid proposes a method of how to carry out a mathematical task. For example, in Proposition 1 of Book 1 Euclid shows how to draw an equilateral triangle (all the sides are the same length and all the angles are equal to 60°), and he then goes on to prove that the triangle is, in fact, equilateral.

ERATOSTHENES (276–195 BC)

It would be wrong to talk too much about prime numbers without mentioning multi-disciplined mathematician Eratosthenes, who hailed from a Greek city in modern-day Libya. He was responsible for many great intellectual endeavours, including calculating the earth's circumference to a surprising degree of accuracy and coining the word 'geography', which means 'drawing the earth' in ancient Greek. Mathematically, Eratosthenes' greatest contribution is the **Sieve of Eratosthenes**.

In their prime

Before we look at the sieve let us first contemplate **prime numbers**: numbers that have only two factors – themselves and 1. Hence 13 is a prime number because 1 and 13 are the only numbers that divide into it without leaving a remainder; 9 is not prime, because it can be divided by 1, 3 and 9, which means it has three factors. 1 is also not a prime number because it has only one factor.

Prime numbers are important for two reasons:

1. Any whole number or **integer** greater than 1 can be written as a chain of multiplied prime numbers. For example, the numbers between 20 and 30 can be written as follows:

    $20 = 2 \times 2 \times 5$
    $21 = 3 \times 7$
    $22 = 2 \times 11$
    $23 = 23 \text{ (prime)}$
    $24 = 2 \times 2 \times 2 \times 3$
    $25 = 5 \times 5$
    $26 = 2 \times 13$
    $27 = 3 \times 3 \times 3$
    $28 = 2 \times 2 \times 7$
    $29 = 29 \text{ (prime)}$
    $30 = 2 \times 3 \times 5$

There is only one way of doing this for each number so it seems to me, at least, that primes are the equivalent of DNA for numbers.

Fundamentals

The idea that any whole number greater than I can be expressed as the unique product of a chain of multiplied prime numbers is called the **fundamental theorem of arithmetic**.

2. They are very mysterious – there is no pattern to prime numbers, and there is no formula that will produce them. To this day the nature of prime numbers is still under intensive study by mathematicians.

Eratosthenes' sieve works using a very simple principle to help find prime numbers up to a certain limit. The number 2 is the first prime. Anything that 2 goes into cannot be prime, because it would then have 2 as a factor as well as itself and 1.

If we set ourselves a limit of 100, we could highlight 2 as a prime and then eliminate all the numbers that have 2 as a factor: 4, 6, 8, etc. up to 100. If we use a grid we can shade them in to generate a pattern:

1	2	3	4	5	6	7	8	9	10
11	12	13	14	15	16	17	18	19	20
21	22	23	24	25	26	27	28	29	30
31	32	33	34	35	36	37	38	39	40
41	42	43	44	45	46	47	48	49	50
51	52	53	54	55	56	57	58	59	60
61	62	63	64	65	66	67	68	69	70
71	72	73	74	75	76	77	78	79	80
81	82	83	84	85	86	87	88	89	90
91	92	93	94	95	96	97	98	99	100

You don't even need to be brilliant at your two-times table to do this – you could just count on 2 each time and shade in each square you land on.

After you've shaded in all the multiples of 2 you move on to the next unshaded number, which also happens to be the next prime number: 3. We highlight 3 as a prime and then eliminate all the multiples of 3, some of which have already been eliminated in the first round. The next unshaded number is 5, which again is also prime. As before, highlight that and then eliminate the multiples of 5.

As you move along, the next unshaded number must be prime because none of the prime numbers that went before it could go into it. If you keep on repeating this process eventually you'll have a completed sieve. Turn to page 55 to see what this looks like.

To Infinity and Beyond

Euclid's theorem demonstrated that there are infinitely many prime numbers. We know that any number can be made by multiplying a chain of prime numbers together; thanks to our sieve, we also now know all of the prime numbers under 100. How can we be sure there are more? Let's use the Sieve of Eratosthenes to investigate.

If you multiply all the primes together you generate a number. This next number will either be a prime number or it won't. If the next number in sequence is a prime then we have a new prime number.

However, if the next number isn't prime there must be a prime number that we don't already know of that goes into making it – therefore there's another prime number somewhere.

So, whatever happens, we either have a new prime number or know there is an unknown prime number that is less than or more than our number. No matter how large we make the sieve, there is always another prime number that is not on it, therefore there must be an infinite number of primes.

1	2	3	4	5	6	7	8	9	10
11	12	13	14	15	16	17	18	19	20
21	22	23	24	25	26	27	28	29	30
31	32	33	34	35	36	37	38	39	40
41	42	43	44	45	46	47	48	49	50
51	52	53	54	55	56	57	58	59	60
61	62	63	64	65	66	67	68	69	70
71	72	73	74	75	76	77	78	79	80
81	82	83	84	85	86	87	88	89	90
91	92	93	94	95	96	97	98	99	100

You can make the sieve as big as you like in order to work out higher and higher prime numbers. There are no hard calculations to do, but it is quite a tedious process – something a Greek mathematician would probably have left to an educated slave.

ARCHIMEDES (287–212 BC)

Archimedes was a friend of Eratosthenes and he hailed from the city of Syracuse in present-day Sicily. He was famous as a scientist and engineer: he invented the Archimedes screw for pumping liquids and raising grain, which is still in use today. Archimedes is also said to have defended Syracuse from Roman warships by directing an intense ray of light from the sun towards the approaching soldiers, setting their vessels alight.

From straight to circular

Archimedes' contributions to mathematics are no less impressive, even if they are less well known. He worked out a value for π by noting that, as a polygon accrues more sides, it gets closer and closer to becoming a circle. π is defined as a circle's circumference divided by its diameter. It is hard to measure the curved edge of a circle accurately, but easy to measure the straight sides of a polygon to find the perimeter. By approximating a circle as a polygon with a certain number of sides, Archimedes was able to find a value for π by dividing the polygon's perimeter by the distance across the polygon. Archimedes performed this calculation with a polygon that had up to ninety-six sides. During his investigations he came up with a value of between 3.143 and 3.140 for π, which is pretty close to its actual value: 3.1415...

Archimedes hit upon an important idea with this **method of exhaustion** – the idea that if an approximation is performed accurately enough it becomes indistinguishable from the true answer. This idea has been used in many other areas of mathematics, perhaps most noticeably in the calculus of Newton (see page 115) and Leibniz (see page 119) almost 2,000 years later.

Archimedes proved other important results, including that the area of a circle is π multiplied by the radius squared. He also proved that the volume of a sphere is ⅔ the volume of the cylinder that it is able to fit into. Archimedes was so pleased with his discovery he had a sculpture of the sphere and cylinder erected on his tomb.

Lasting legacy

Archimedes died at the hands of a Roman soldier while working at his desk. Legend suggests Archimedes was so absorbed in his work he failed to respond to the soldier's orders that he come with him. Insulted, the soldier killed Archimedes, and presumably faced the wrath of his commanding officer, who would have regarded the slain intellectual as a highly valuable scientific asset.

With the death of Archimedes we come to the end of ancient Greece, when its territories were consolidated into the emerging Roman Empire. The mathematical legacy of the Greeks is long lasting and most people today will have encountered the discoveries made by many of the mathematicians mentioned in this chapter. I think the ancient Greeks' greatest contribution was to invent mathematics as a rich and diverse subject, moving it beyond the basic necessity of numeracy and arithmetic, the functional tools of an economy. They created a subject that would become the language of science and which would eventually allow humanity to create scientific ideas from first principles, basing discoveries on a concept rather than from fudging equations and formulae to match observations. Without this mode of thinking Sir Isaac Newton would have been unable to conduct much of his pioneering work.

The Romans

The Greek mainland was conquered by the Romans in 146 BC, and the empire reached its zenith 200 years later, occupying a vast area that covered the entirety of the Mediterranean on all sides.

A PRACTICAL PEOPLE

Discipline was a central aspect of Roman life, which extended to its education system. The wealthier young Romans were taught basic arithmetic, most likely at home, but the main thrust of their education was to understand the workings of their own society. Oration was seen as the pinnacle of education, along with physical training for boys, who would go on to do military service, and home economics for girls, who were in charge of running their homes.

In terms of higher mathematics, it appears that very little was taught to the Romans when compared to their Greek predecessors. The Romans were a far more practical people, focusing their attentions on developments in engineering and medicine; practicality is not the best mindset for exploring mathematics for its own sake.

A spanner in the works

The Roman number system, inherited from the Greeks, didn't help matters. Roman numerals rely on the position they occupy within a string of letters, which makes it very difficult to use them in arithmetic.

The basic Roman numerals are:

I: one
V: five
X: ten
L: fifty
C: one hundred
D: five hundred
M: one thousand

The Romans wrote their numbers with the largest starting from the left. Therefore, in order to read a Roman numeral you have to add up the numbers from left to right. For example:

MMMDCLXVII would be 1000+1000+1000+500+100 +50+10+5+1+1 = 3667

However, the Romans devised a useful shortcut for using when the value of a number was close to the value of the next letter. The method involved putting a letter out of sequence, which indicated it should be subtracted from the next letter in sequence.

For example, in longhand the number 999 should be written DCCCCLXXXXVIIII, but with the shortcut it could be written as IM. However, there seemed to be no written rules, and the Romans, it seems, didn't like having an I before an M or a C if they could avoid it. Therefore, 999 would more likely have be written as CMXCIX which gives $(1,000 - 100) + (100 - 10) + (10 - 1) = 900 + 90 + 9 = 999$. Needless to say, having more than one way to write a number did not make life easy!

Alexandria

The Romans subsumed the old Greek Empire and, as such, the Greek mathematical tradition continued. It focused in Alexandria, Egypt, a remarkable centre of learning that had been founded in 331 BC by the leader of the Greeks, Alexander the Great, as he conquered his way east across Europe and Asia.

HERO (10–70 AD)

An Alexandrian scientist and mathematician, Hero is most famous for detailing a primitive steam engine, and for perhaps being the first person to harness wind power on land with the aid of a windmill.

Hero also made two significant contributions to mathematics:

1. He came up with a formula for working out the **area of a triangle** that only requires the lengths of the sides of the triangle.

2. He devised a way of working out **square roots**: a number that when multiplied by itself gives a specific quantity.

Hero's formula

There are many ways to work out the area of a triangle. Most of us were taught at school that:

> **area of triangle = ¹/₂ × base × height**

For this formula you need to choose which side is the base and then work out the height of the triangle, which, if it's non-right-angled, may not be one of the other two sides:

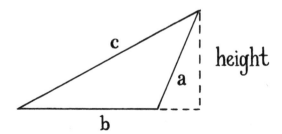

Hero's formula removed the need both to choose a base and to measure the height, although perhaps at the expense of simplicity:

$$\text{area of triangle with sides of length a, b and c}$$
$$= \tfrac{1}{4} \times \sqrt{[(a^2 + b^2 + c^2)^2 - 2(a^4 + b^4 + c^4)]}$$

The root of the problem

Hero's method for working out square roots involved using a formula to generate a new value; this new value would then be put back into the formula and the process would be repeated a number of times with the answer getting closer to the true value.

This technique is called **iteration** – another important development in mathematics. For example, if you wanted to work out the square root of 2, which, as we saw earlier, is an irrational number – one which cannot be written as a fraction and whose decimal goes on for ever without repeating (see page 42) – Hero's method would work like this:

$$\text{new value} = \tfrac{1}{2} \times (\text{old value} + R \div \text{old value})$$

where R is the number you want to know the square root of. The first time you use the formula there is no 'old value', so you have to take a guess. The square root of 2 must be between 1 and 2, because $1 \times 1 = 1$ and $2 \times 2 = 4$ and 2 lies between 1 and 4. Let's opt for the middle value, 1.5, and see what happens:

new value = $\frac{1}{2}$ × (1.5 + 2 / 1.5) = 1.41666666...

You can now repeat this process using 1.41666 as your old value:

new value = $\frac{1}{2}$ × (1.41666 + 2 / 1.41666) = 1.414215686

new value = $\frac{1}{2}$ × (1.414215686 + 2 / 1.414215686) = 1.41423562

new value = $\frac{1}{2}$ × (1.41423562 + 2 / 1.41423562) = 1.41423562

At this point you should notice that the old value and new value are the same, so our work here is done – and this is indeed the square root of 2, accurate to eight decimal places.

If you wanted to work out the square root of another number you would start with a different R. It's important to note that if you make R a negative number the formula does not work. For example, if you make R = -2 and have 1 as your first guess you get:

new value = $\frac{1}{2} \times (1 - \frac{2}{1})$ = -0.5

If you repeat as before you get: 1.75
0.3035714286
-3.142331933
-1.252930967
0.1716630854

This process continues for ever without ever settling on a value. Why? Because negative numbers cannot have a square root – a negative number multiplied by a negative number always gives a positive answer. Hence the formula is searching for something that does not exist!

Hero did, however, postulate that it could be possible for a negative number to have a square root, if you use a bit of imagination (see page 128).

DIOPHANTUS (c. 200–c. 284 AD)

A resident of Alexandria from *c.* 250 AD, Diophantus is sometimes referred to as the 'Father of Algebra' because of his contribution to solving equations. While today thoughts of algebra conjure up a process of replacing numbers with letters, Diophantus did not adhere to this principle. Before true symbolic algebra was invented, mathematicians were forced to write out equations longhand.

These days it's very easy to write simple algebraic equations, such as: $3a + 4a^2$. However, Diophantus' method would have been far more laborious, involving something along the lines of: 'three multiplied by the unknown number added to four times the unknown number multiplied by itself.' This made solving equations a tricky process, both in terms of writing and reading them.

An imaginary triangle

Diophantus was interested in Pythagoras' theorem. He noticed something strange when he tried to work out the sides of a right-angled triangle with a perimeter of 12 and an area of 7. It produced an equation that could not be solved, indicating a triangle with those specific dimensions cannot exist. However, Diophantus remarked that if negative numbers could have square roots he would be able to solve the equation – which would mean the triangle would then exist. Much later, these numbers were called **imaginary numbers** (see box on page 128), because in order to get round the problem you have to imagine that there is a number, represented by the symbol 'i', that is the square root of -1.

Triple the fun

His interest in Pythagoras' theorem also sparked another mathematical mystery that would take hundreds of years to solve. Diophantus was interested in **Pythagorean triples**, which are solutions to the theorem that are whole numbers. For example:

$$3^2 + 4^2 = 5^2$$
$$5^2 + 12^2 = 13^2$$
$$8^2 + 15^2 = 17^2$$

In his great work *Arithmetica*, Diophantus included instructions on how to find such numbers. In 1637, French mathematician Pierre de Fermat wrote in the margin of his copy of *Arithmetica* that it was not possible to find the Pythagorean triples where the numbers were raised to any power other than 2. He finished with a tantalizing comment that was to tease mathematicians for years to come: 'I have discovered a truly marvelous proof of this, which this margin is too narrow to contain.'

These innocuous words started a 350-year challenge to solve what became known as **Fermat's last theorem**.

Unravel the Riddle

Although we know very little about Diophantus' life, a charming riddle, sometimes known as 'Diophantus' Epitaph', associated with him provides a brief overview of his days on this earth. The riddle was first noticed in a puzzle book by the Greek philosopher Metrodorus some time in the sixth century AD.

'Here lies Diophantus,' the wonder behold.
Through art algebraic, the stone tells how old:
'God gave him his boyhood one-sixth of his life,
One twelfth more as youth while whiskers grew rife;
And then yet one-seventh ere marriage begun;
In five years there came a bouncing new son.
Alas, the dear child of master and sage
After attaining half the measure of his father's life
chill fate took him. After consoling his fate by the
science of numbers for four years, he ended his life.'

Can you work out how old Diophantus was when
he died?

Hypatia (*c.* 370–415 AD)

The Alexandrian mathematician, philosopher and astronomer Hypatia was the daughter of Theon, a mathematician who produced an edition of Euclid's *Elements*. He educated his daughter in the same way as his sons, which exposed Hypatia to the rich philosophical heritage of her Greek ancestors.

Hypatia was a teacher specializing in the philosophies of Plato and Aristotle, and as part of this she developed her own ideas in mathematics, physics and astronomy. She edited her father's editions of Euclid's and Diophantus' works, using her teacher's eye to help the reader understand the more difficult sections.

Hypatia is widely considered to have been the first woman to make contributions to mathematics and science, although few of her original works survive. She dressed in scholar's robes rather than in female dress, and chose to navigate the city unaccompanied, often driving her own chariot, which at the time was considered very unladylike. Hypatia also stood for what by then the Christian Romans considered to be a pagan religion. Her lectures, which were open to all comers, regardless of race or religion, were targeted by Christians and led to riots. This discrimination reached an inevitably bloody conclusion, and in March AD 415 Hypatia was brutally attacked and murdered by a Christian mob.

THE END OF THE ROMANS

The Roman Empire began to disintegrate in *c.* AD 380. In the absence of the sizeable bureaucratic machine and enforced discipline the Romans had instilled, Western Europe entered what is sometimes referred to as the Dark Ages: a period when little intellectual development occurred. Allegedly.

Eastern Mathematics

$$y = ax^2 + bx + c$$

$$\frac{-b \pm \sqrt{b^2 - 4ac}}{2a}$$

The history of mathematics was not confined to Europe. China, India and the countries of the Middle East each have a tradition rich in the subject, and the flow of mathematical knowledge was, generally speaking, from East to West. As Europe found itself plunged into the Dark Ages, mathematical discoveries in the East ensured the subject continued to go from strength to strength.

CHINESE MATHEMATICS

Chinese history is populated by dynasties – a succession of ruling families, each of whom prioritized eradicating all evidence of the previous incumbent. As such, many important Chinese mathematical works and artefacts have been lost over time.

Much of what we know about Chinese mathematics is

attributed to a scholar and bureaucrat called Qin Jiushao (1202–61). He wrote a book called *Mathematical Treatise in Nine Sections* that discusses practical mathematics in a variety of fields relevant to government officials. Jiushao's book also contains a detailed history of Chinese mathematics, and sheds light on the country's mathematicians and their advances in the field.

It All Adds Up

Chinese numbers were based on a system of **counting rods**: short sticks that, when placed in certain arrangements, denoted various numbers in a decimal system. Their written numerals were simply drawings of the arrangement of these sticks.

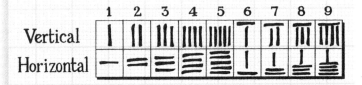

In c. AD 700 the Chinese borrowed the concept of zero from India (see page 75), which means they were one of the first cultures to have a fully fledged decimal number system.

Predicting the future

The *I Ching* (*Book of Changes*) is a famous Chinese text that dates from, at the very least, *c.* 1000 BC, and quite possibly before then. The text allows you to divine your future using **trigrams** and **hexagrams**, both of which have their origins in mathematics.

A trigram is a stack of three horizontal lines, which can be either *yang* (solid) or *yin* (broken). It is possible to make eight different trigrams using this system, and each trigram has various attributed meanings, including the Chinese elements: earth, mountain, thunder, water, lake, wind, heaven and fire.

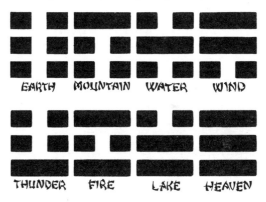

EARTH   MOUNTAIN   WATER   WIND

THUNDER   FIRE   LAKE   HEAVEN

Two trigrams could be combined to make a hexagram, and there are 64 (8×8) possible hexagrams to be made from the eight trigrams – which could then be used to predict your future. Soothsayers would need to be familiar with the interpretations of each trigram and hexagram in order to use them to build up your reading.

The German philosopher Gottfried Leibniz (see page 119) was intrigued by Chinese philosophy and noticed that the trigrams and hexagrams of the *I Ching* can be written as **binary numbers** – a system of numbers that has 2 rather than 10 as its base – if the *yang* is seen as 1 and the *yin* as 0.

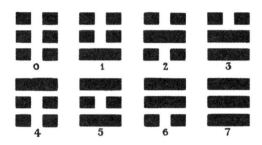

Leaps and bounds

Zu Chongzhi (AD 429–500) was a Chinese astronomer and mathematician whose discoveries lay far ahead of his time. Chongzhi calculated various astronomical constants to extremely high degrees of precision; he also worked out independently a value for π using Archimedes' method of exhaustion on a polygon with over 12,000 sides. His answer gave a working value of 355/133, which is accurate to six decimal places (see page 156). Europe would not achieve this level of precision for over 1,000 years.

The *Nine Chapters on the Mathematical Art* is one of the oldest and most important Chinese mathematical works, compiled over the centuries up to *c.* AD 100. It gives us a very good idea of the state of Chinese mathematics that existed at approximately the same time as Greek civilization. The chapters covered the following topics:

1. Areas of fields
2. Exchange rates and prices
3. Proportions
4. Division; square and cube roots; area of a circle and volume of a sphere
5. Volumes of other solids
6. Taxation
7. Solving equations
8. Simultaneous equations
9. Pythagoras' theorem

While the Chinese may have been more concerned than the Greeks with practical matters, we can see that their development in mathematics was on a par.

Court eunuch Jia Xian (AD 1010–70) is credited with being the first individual to investigate what later became known in Europe as **Pascal's triangle** (see page 82). Chinese mathematician Yang Hui (AD 1238–98) published Xian's findings in 1261, four hundred years before Pascal's discoveries would be revealed. Xian was also interested in what we call **magic squares**, which had long fascinated Chinese mathematicians. A magic square is a square of numbers in which all the rows, columns and diagonals add up to a particular number. For example:

$$
\begin{array}{ccc}
4 & 9 & 2 \\
3 & 5 & 7 \\
8 & 1 & 6
\end{array}
$$

In the example above, each row, column and diagonal has a

sum of 15. This particular magic square is known as the *Lo Shu* square because of its connection to a legend in which the River Lo floods and a magic turtle carries the magic square on its shell to aid the afflicted people.

The Chinese Abacus

At some point in c. AD 1000 the Chinese began to use the *suanpan* (Chinese abacus) in favour of the counting rods, although the *suanpan* had been around for some time and may have influenced the abacus in the West.

This particular abacus has counters suspended on rods, which are layered on two decks. On the lower deck there are five beads per rod, each of which is worth one; on the top deck there are two beads, each worth five. Each rod represents a decimal column (unit, tens, hundreds, etc.) and pushing the beads towards the separator in the middle signifies the number. For example, 123,456 would look something like this:

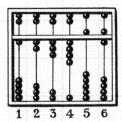

The extra bead on each deck can be used for performing calculations.

INDIAN MATHEMATICS

In 1920 archaeological excavations in north-western India unearthed the Indus Valley civilization, which existed from *c.* 3500 BC to 2000 BC. These Bronze Age settlements, contemporaneous with the first urban areas in Egypt and Mesopotamia, indicated the ancient Indians had a good understanding of basic mathematical concepts, and possessed a standardized system of weights and measures.

Ancient Indian religious texts also contain evidence of mathematical knowledge; in Hinduism, mathematics, astronomy and astrology were considered to be in the same field, and they each had important religious implications. It was a religious requirement that all altars should occupy the same amount of floor space, even if they weren't the same shape or used different configurations of bricks – all of which required a good knowledge of geometry. Texts from 700 BC show the ancient Indians possessed knowledge of Pythagoras' theorem, irrational numbers and methods for calculating them.

Astronomical discoveries

Brahmagupta (AD 598–668), an astronomer, was the first person to treat zero as a number. The Hindu numeral system, predecessor to the Hindu-Arabic numeral system that we use today (see page 79), developed over time and was fully established by the end of the first millennium AD. Up until Brahmagupta's treatment of zero as a number, it had been used merely as a placeholder within various number systems in order to show a gap. Brahmagupta, however, thought of 0 simply as a whole number or **integer** that lies between 1 and -1. He wrote down rules for its use in arithmetic, alongside rules for using negative numbers.

Useful Functions

Aryabhata (AD 475–550) was an astronomer who is credited with being the first person to introduce **trigonometry**, which we use to work out lengths and angles in triangles, and the concept of the **sine**, **cosine** and **tangent** functions.

Brahmagupta recognized that an equation could have a negative solution and, as a result, that any positive integer would have a positive and negative square root. For example, the square roots of 36 are 6 and -6, because, as Brahmagupta himself stated, a negative multiplied by a negative gives a positive.

Brahmagupta is also famous for developing **Brahmagupta's formula**, which tells us the area of a **cyclic quadrilateral** – a four-sided shape, the corners of which lie on a circle:

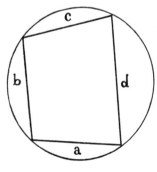

Modern Indian Mathematics

Srinvasa Ramanujan (1887–1920) was an Indian mathematical genius. After dropping out of university, he became an accounting clerk at a government office, from where he sent papers to various British mathematicians for consideration. The English mathematician Godfrey Hardy (1877–1947) recognized Ramanujan's genius and arranged for him to have a research post at the University of Madras.

In 1914 Ramanujan joined Hardy at Cambridge University and remained in England for five years, in which time he became one of the youngest ever members of the Royal Society, had work published and finally gained a degree. However, Ramanujan was often ill.

During one bout of illness, Hardy visited him and mentioned that the number of his taxi, 1729, was 'rather dull'. Ramanujan replied instantly that 1,729 was the lowest number that could be written as the sum of two cubes in two different ways, and as such, was actually quite interesting:

$$1^3 + 12^3 = 1 + 1,728 = 1,729$$
$$9^3 + 10^3 = 729 + 1,000 = 1,729$$

There are lower numbers that can be written as the sum of two cubes, but 1,729 is the lowest number that can be written like this in two ways, and Ramanujan's instant recognition of this was nothing short of miraculous.

In his short life Ramanujan came up with nearly 4,000 theorems, equations and identities that still inspire mathematical research to this day.

If you find half the perimeter of the quadrilateral (let's call it 's') then the area of the shape can be found using Brahmagupta's formula:

$$\sqrt{(s-a)(s-b)(s-c)(s-d)}$$

Although the Indians were clearly excellent mathematicians, when the British began to take control of the country in the 1700s they assumed the backward pagan Hindus had nothing of worth to contribute beyond vast natural resources and cheap labour. It has only been in the last hundred years that we have come to appreciate the mathematical heritage of the subcontinent.

ISLAMIC MATHEMATICS

Mohammed, the founder of Islam, was born in AD 570. In the two centuries following Mohammed's birth the Islamic Empire came to dominate all of the Middle East, Central Asia, North Africa and what would become Spain and Portugal. This Islamic Golden Age saw much important mathematical progress emerge from the countries in the empire, while Europe remained still in its Dark Ages.

The Islamic religion itself is particularly open to the idea of science, which contrasted strongly with the ideas prevalent in medieval Europe, where it was often considered heretical to question or investigate the workings of a world made by God.

The Islamic Empire too was committed to gathering the knowledge of the ancient world. Texts in Classical Greek and Latin, ancient Egyptian, Mesopotamian, Indian, Chinese and Persian were all translated by Islamic scholars, broadening their availability to the empire's scientists and mathematicians.

AL-KHWARIZMI (C. 790–C. 850)

Mathematician Al-Khwarizmi hailed from an area situated in present-day Uzbekistan, and he is credited with providing several significant contributions to mathematics. Although some of his original works have survived, he is familiar to us through editions of his work translated into Latin for use later in Europe.

The new number system

One of Al-Khwarizmi's significant legacies is what is now known as the Hindu-Arabic numeral system, which we still use to this day. Derived from his *Book of Addition and Subtraction According to the Hindu Calculation*, Al-Khwarizmi's system of numbers, developed over time in India from *c.* 300 BC and passed through into Persia, revolutionized arithmetic.

Up to this point, no culture had a system of numerals with which it was really possible to use in arithmetic. Numbers would always be converted into letters or symbols (either mentally or using counters, abacuses or other such tools), the calculation performed and the result reconverted back into numerals. Lots of symbols were often needed to show a number, many of which were difficult to decipher at a glance.

The Hindu-Arabic system contains just ten symbols – 0 1 2 3 4 5 6 7 8 9 – that could be used to write any number. It is important to note that these symbols were exactly that – they were not associated with the value they represented through stripes or dots. The zero (from the Arabic *zifer*, meaning 'empty') meant that the symbols could have a different value depending on where they were positioned in the number – which freed people of the difficulty the Mesopotamians had faced. Today, the concept of place-value is taken for granted. But the idea that the 8 in 80 is worth eight tens, and yet could be used, with the help of those friendly zeros, to also mean 800 or 8 million was revolutionary at the time. In fact, some European scholars were deeply suspicious of this heathen method of calculating, despite its advantages.

In the *Book of Addition and Subtraction According to the Hindu Calculation* Al-Khwarizmi describes how to do arithmetic using these new numbers. His translators referred to him by the Latinized name Algorism. Over time Al-Khwarizmi's methods of calculation became known as **algorithms**, a word still in use today and which refers to a set of instructions to perform a calculation – which is exactly what Al-Khwarizmi provided.

Transforming Mathematics

Al-Khwarizmi also wrote *The Compendious Book on Calculation by Transformations and Dividing*, which set out to show how to solve different types of **quadratic equations** (equations in which the unknown numbers are squared). 'Transformations' in Arabic is *Al-Jabr*, from which we derive (via Latin) the English term **algebra**. While Al-Khwarizmi himself did not replace unknown numbers with letters, he did pave the way for this to happen.

OMAR KHAYYÁM (1048–1131)

Persian scholar Omar Khayyám is best known for his *The Rubaiyat of Omar Khayyám,* a selection of poems that were later translated into English in the nineteenth century by the poet Edward Fitzgerald. Multi-talented, Khayyám spent a great proportion of his life as a court astronomer to a sultan, while also working as a scientist and mathematician.

Khayyám's mathematical works were far-reaching. He expanded on Al-Khwarizmi's earlier work in algebra, and he was one of the first mathematicians to use the replacement of unknown numbers with letters to make solving equations easier. He also devised techniques for solving **cubic equations**, where the unknown term has been cubed. Khayyám's insight enabled him to be one of the first people to connect geometry and algebra, which had until that point been separate disciplines.

Endless possibilities

Khayyám also investigated something now called the **binomial theorem**. This has many applications in mathematics, many of which involve rather tricky algebra. One side product of binomial theorem is something called **Pascal's triangle**, named after the seventeenth-century French mathematician Blaise Pascal, who borrowed the triangle from Khayyám, who in turn borrowed it from the Chinese (see page 73). Unlike the binomial theorem, Pascal's triangle is simple to understand: the number in each cell of a triangle is made by adding together the two numbers above it.

Pascal's triangle is useful because each horizontal row shows us the **binomial coefficients** that the binomial theorem spits out. These can tell us how many combinations of two different things it is possible to have.

For example, imagine you have planted a row of four flower

bulbs. It says on the packet that the flowers can be blue or pink, with an equal chance of having either.

There is one way for you to grow four blues:

BBBB

There are four ways for you to grow three blues and one pink:

BBBP
BBPB
BPBB
PBBB

There are six ways for you to end up with two of each:

BBPP
BPPB
PBBP
PPBB
BPBP
PBPB

There are four ways for you to have three pinks and one blue:

PPPB
PPBP
PBPP
BPPP

And one way for you to have four pinks:

PPPP

If you look across the fourth row of the triangle, it says 1, 4, 6, 4, 1, which corresponds to the number of ways worked out in the example above. Because there is an equal chance of a flower being either pink or blue, you can also see that you're most likely to end up with two of each colour because there are 6 out of 16 total ways this could happen.

A new geometry

Khayyám also wrote a book that tackled Euclid's fifth postulate, which had long rankled a contingent of mathematicians. The fifth postulate Euclid wrote concerned parallel lines, and it is therefore normally referred to as the **parallel postulate**.

Imagine two lines (PQ and RS) with a third (XY) crossing them. Inside PQ and RS we now have four angles, two on each side of the XY: a, b, c and d:

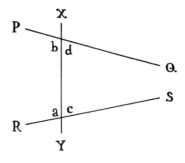

The parallel postulate suggests that if you add the pairs of angles on the same side of XY together (e.g. a+b and c+d) then PQ and RS will cross on the side of the line where the sum of the angles is less than 180°. If the angles on each side add up to 180° then PQ and RS are parallel and therefore will never cross.

Mathematicians, however, have argued over the ages that this postulate is not quite as obvious as Euclid made out. Khayyám was the first to come up with a counter-example, arguing that Euclid's parallel postulate does not always work if the surface you are drawing on is curved. Thus Khayyám instigated the ideas of **elliptical** and **hyperbolic** geometry, a direct challenge to the simple **Euclidean** geometry that had gone before. This kind of thinking would eventually help Albert Einstein to come up with his ideas of space-time and gravity.

The Middle Ages in Europe

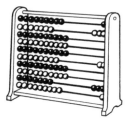

Despite Europe's plunge into the Dark Ages – so-called because it was thought that following the fall of the Roman Empire the continent had reverted back to a barbaric state of tribal warfare and religious fundamentalism – there remained a coterie of individuals intent on pushing the boundaries of mathematics even during these difficult times.

BEDE (672–735)

The Venerable Bede is known more perhaps for his contribution as a historian than for the role he played in the development of mathematics. Bede was a monk living in north-eastern England and his translation of a number of scholarly works into the English of the time helped to spread an enormous amount of knowledge.

Bede's contribution to mathematics began when he attempted to develop a way to calculate accurately when Easter would fall. At the time it was thought to fall on the first Sunday after the first full moon following the spring equinox. Missing Easter mass following the calculation of an incorrect date would have resulted in excommunication, and therefore damnation, so Bede's was no trivial task.

Dating in the Dark Ages

In order to calculate the date of Easter, it was necessary for Bede to rationalize the date of the spring equinox with the lunar calendar. This was a difficult task in itself as the date of the equinox varied because the Julian calendar in use at the time was unreliable. Because the date of the equinox varied each year, and full moons come at alternate twenty-nine- or thirty-day intervals, it meant that there was a nineteen-year cycle of possible dates for Easter. The procedure for calculating the date of Easter has been known as computus (meaning 'computation') ever since.

Once Bede had completed the computus, he decided to sort out dating the rest of history as well. Prior to Bede's endeavours, historians had been dating things in reference to the lifetime of the current emperor or king; for example: 'the Vikings first attacked in the third year of Aethelred's reign.' This method, of course, relied on the reader knowing when Aethelred was around in the first place. Bede decided that it would be far more sensible to date everything occurring either before or after the birth of Jesus Christ. Although not originally Bede's idea – that responsibility lay with Dionysius Exiguus, a south-eastern

European monk active during the sixth century – such was his influence that we have been using AD (*Anno Domini*, Year of the Lord) and BC (Before Christ) ever since.

Finger Talk

Bede also wrote a book called *On Counting and Speaking With the Fingers*, which allowed the reader to use hand signals for numbers into the millions – a super-sized version of the systems we saw Stone Age cultures using. Again, such was his influence, people were still referencing Bede's book 1,000 years later.

ALCUIN OF YORK (730–804)

A gifted poet, scholar, teacher and mathematician, Alcuin of York began his academic life under the instruction of Archbishop Ecgbert of York, who in turn had been tutored by Bede. Alcuin's main mathematical work was a textbook for students titled *Propositiones ad acuendos juvenes* (*Problems to Sharpen the Young*). The book contains many word-based logic puzzles, a few of which have become quite famous, including the following two river-crossing problems.

Heavy load

The first problem relates to a man trying to cross a river with a wolf, a goat and a cabbage. The man's boat is very small and he can only fit one thing in the boat with him at a time. However, if he leaves the goat and the wolf together, the wolf will eat the goat. If he leaves the goat and the cabbage together, the goat will eat the cabbage. How does he get them all safely across the river?

Answer: this is a good medieval example of lateral thinking. Clearly, on his first run the man can only take the goat across the river. On his second trip he brings the wolf across but *takes the goat back with him*; he then leaves the goat there and takes the cabbage across, and then makes a final trip for the well-travelled goat.

Family matters

In the second problem, a couple, who are of equal weight, have two children, each of whom weighs half the weight of one of the adults. All four people need to cross a river, but their boat will only hold the weight of one adult. How can they cross in safety?

Answer: the children cross the river in the boat. One child stays on the far bank while the other child returns. Dad crosses to the far bank and the child returns with the boat to be with the mum and the other child. The two children cross again and one remains on the far side with Dad. The other child returns to be with Mum. Mum crosses to the far side, and the child with the dad returns to collect the other child to reunite the family.

An Education

In 781 Alcuin joined the court of Charlemagne, the King of the Franks, where his skills as a teacher were in great demand. While there, Alcuin introduced the **trivium** and **quadrivium**, which he had encountered during his time in York.

During medieval times only seven subjects were taught in schools and universities. The trivium (Latin for 'three roads') comprised logic, grammar and rhetoric. Logic was seen as the way to organize one's thinking, grammar the way to express these thoughts without confusion, and rhetoric was the way to persuade others that your correctly expressed thoughts were worth listening to.

After graduating in the trivium, worthy students could attempt the quadrivium (Latin for 'four ways'): geometry, arithmetic, astronomy and music.

The trivium was the equivalent of an undergraduate course and the quadrivium a Master's degree. Succeeding in these courses of study gave access to the Doctorates, either of Philosophy or Theology.

GERBERT D'AURILLAC, POPE SYLVESTER II (946–1003)

Born in France, D'Aurillac joined a monastery during his teenage years, from where he was sent to Spain for further education. Under significant Arabic influence, Spain exposed D'Aurillac to the wonderful discoveries of the Islamic mathematicians. D'Aurillac carved a name for himself as an excellent teacher and was taken on as a royal tutor. His political career soon followed, culminating in him becoming the first French Pope in the year 999.

In his elevated position, D'Aurillac introduced the Hindu-Arabic number system to Europe (see page 79), although it did not immediately become widely accepted. He was also responsible for reintroducing the abacus, which had not been used since Roman times but which soon became commonplace.

LEONARDO OF PISA (FIBONACCI) (c. 1170–1250)

The son of an Italian trader, Fibonacci lived near Algiers in North Africa, where he gained his first taste for Arabic mathematics. He travelled widely around the Islamic world to further his learning and published a seminal book on his findings called

Liber Abaci (*Book of the Abacus*). Fibonacci's approach to writing the book showed a keen head for business – not only did he expound the advantages of the Hindu-Arabic number system, he also applied it directly to banking and accounting. Fibonacci's book became very popular among medieval European scholars and businessmen, and his success earned him the patronage of the Holy Roman Emperor. A triumph, Fibonacci was then able to continue his mathematical work in the fields of geometry and trigonometry.

Fibonacci's name is well known for the sequence of numbers named in his honour. The sequence derived from, of all things, a problem about rabbits that he posed in his *Liber Abaci*.

At it like rabbits

Fibonacci numbers were known to Hindu mathematicians long before Fibonacci encountered them, but, much like Blaise Pascal (see page 82), Fibonacci became eponymous with the sequence after it appeared in his book. In his problem, Fibonacci considers the growth of a rabbit population in a field. Fibonacci conjectured rabbits could start mating after they'd reached the age of one month, and could reproduce every month thereafter. Therefore, if you start with one pair of newborn rabbits (one male and one female) in a field, how many pairs will you have in one year (if each female rabbit continues to breed one male and one female)?

MONTH   N.º of PAIRS

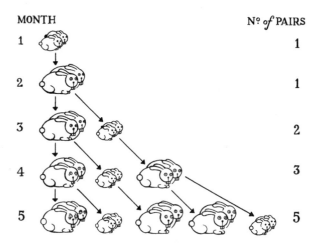

1   1

2   1

3   2

4   3

5   5

You can see the pattern emerging in the right column: 1, 1, 2, 3, 5. You can calculate the next number in the sequence by adding the previous two numbers in the sequence together. So, next month there will be 3 + 5 = 8 pairs in the field. If you continue with this pattern until there are 13 terms in the sequence (which takes you to the end of the twelfth month of the year), you get the following sequence of numbers: 1, 1, 2, 3, 5, 8, 13, 21, 34, 55, 89, 144, 233. Therefore there will be 233 pairs of rabbits at the end of a highly sexed, not to mention very incestuous, year.

All Part of Nature

Although the rabbit example is biologically inaccurate, the Fibonacci numbers do crop up in all manner of natural settings:

- The number of petals on some flowers form part of the Fibonacci number sequence.

- Plant shoots often split in such a way that the number of stems follows a Fibonacci pattern.

- The scales on a pineapple make three spirals, each of which contains a Fibonacci number of scales.

A SYMBOLIC SHIFT

Today, even the least mathematically minded schoolchild understands the four symbols we use for arithmetic: + - × ÷, and the sign we use to show our answers, =. Before the invention of these shorthand ways of writing, the words were written out in full, which made following a mathematical treatise even more cumbersome than it is now.

Geoffrey Chaucer (1343–1400)

One of the great poets of the Middle Ages, Geoffrey Chaucer is not normally associated with the disciplines of science and mathematics. Chaucer, however, also had a sideline in astronomy and alchemy, the latter of which sought to discover the philosopher's stone, the means by which base metals could be turned into gold.

As part of these activities Chaucer became an expert at using a device called an astrolabe, a circular disc that allows you to find certain celestial bodies in the night sky for any given latitude. Having detected his son Lewis' interest in science from an early age, Chaucer wrote *A Treatise on the Astrolabe* in his honour. Understandably, perhaps, it is a rather dry text, despite Chaucer's attempts to enliven the subject by writing the book in verse.

The words *plus* and *minus* are, respectively, Latin for 'more' and 'less'. In medieval times the letters 'p' and 'm' were used to denote these two actions, until German mathematician Johannes Widmann first used the + and - symbols in his 1489 work *Nimble and Neat Calculation in All Trades*.

Next came the equals sign, =, first used in Welsh physician and mathematician Robert Recorde's catchily named book *The Whetstone of Witte* (1557) – one of the first books on algebra to be published in Britain. In *The Whetstone of Witte*, Recorde states his intention to use symbols 'to avoid the tediouse repetition of these words'.

The multiplication symbol, ×, came later on in Englishman William Oughtred's book *The Key to Mathematics*, which was published in 1631. John Wallis, chief cryptographer for Parliament, first used the ouroboros symbol, ∞, to mean infinity, in his 1665 book *De sectionibus conicis*, in which he considers cones and planes intersecting to form curves (which today is referred to as **conic sections**).

The division sign, ÷, is technically called an *obelus*, and it was first used in Swiss mathematician Johann Rahn's book *Teutsche Algebra* in 1659. Ever concise, Rahn was also the first to use '∴' to mean 'therefore'.

The Dark Ages, it seems, weren't quite so barren after all. The slow diffusion of mathematical knowledge from the East allowed Europeans to catch up gradually with their Islamic counterparts. And what happened next allowed the Europeans to take the lead...

The Renaissance Onwards

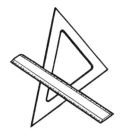

The Renaissance began in Italy as early as the twelfth century and witnessed great advances in all fields of intellectual endeavour. It sparked a new-found interest in the culture of the Classical civilizations, which made an appreciation for – not to mention an investment in – science, culture and philosophy the done thing among the wealthy patrons of fourteenth-century Europe.

The initial epicentre of the Renaissance was Florence, Italy, where a wealthy merchant family, the Medicis, became sponsors of art and culture. One artist who benefited from their philanthropy was Leonardo da Vinci.

Leonardo da Vinci (1452–1519)

Legendary for his talent in almost every field of intellectual pursuit, da Vinci was equally adept in the arts and the sciences. His superior ability and imagination enabled him both to paint the *Mona Lisa* and to invent flying machines, among other extraordinary feats.

Perfectly proportioned

Da Vinci was also a keen anatomist, perhaps out of a desire to bring an element of realism to his art. He was very interested in the relative proportions of the human body, and his famous drawing of the *Vitruvian Man* demonstrates his understanding.

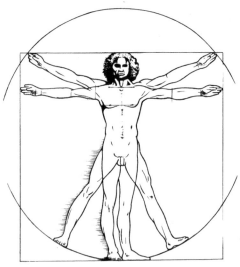

The name of the picture harks back to a Roman architect called Vitruvius. He believed the proportions of the human body are naturally pleasing to the eye, which led him to design his own buildings along similar proportions. Vitruvius considered the navel to be the natural centre of a man's body. He believed a square and a circle drawn over an image of a man with his legs and arms outstretched would represent the natural proportions of the body. Many artists tried to draw human figures that adhered to Vitruvius' proportions, but all looked somehow misshapen. Da Vinci, however, discovered the correct drawing could be made if the centre of the circle and the centre of the square are positioned differently.

A Helping Hand

Although da Vinci himself was not a trained mathematician, he did spend time receiving instruction from Luca Pacioli, a highly regarded maths teacher and accountant. Da Vinci created many drawings of solids for one of Pacioli's books, and his technical expertise with perspective helped to make the diagrams clear.

NICOLAUS COPERNICUS (1473–1543)

A Polish astronomer, Nicolaus Copernicus was one of the first to propose the heliocentric model of the universe: the earth is not the centre of the universe; it does, in fact, orbit the sun. While this was not a mathematical discovery in itself, the way in which Copernicus devised his theory had significant implications for mathematics and science.

The march of science

According to the Bible, the earth was the centre of the universe, which was a perfectly reasonable, if slightly self-important, assumption to have made. After all, both the sun and the moon appear to orbit the earth every day; indeed, all other objects in the night sky appear to do the same.

However, an exploration into the world of astronomy soon revealed problems with this assumption. For example, there are times when the planets appear to move in reverse, which could not be explained if the earth is stationary.

Scientists at the time worked empirically, which means they made observations of phenomena and then came up with an explanation to fit what had been observed. But Copernicus did something that was considered very backward by scientists at the time – he first came up with a theory about how the solar system might work and then tested it against observations, using mathematics as his main tool.

While Copernicus's heliocentric model did not cause much of a stir at the time, his way of working theoretically was one of the first examples of a new way of conducting modern scientific methods.

JOHN NAPIER (1550–1617)

A Scottish nobleman, John Napier was responsible for inventing a new kind of abacus called **Napier's bones**. He also discovered **logarithms**.

Napier's table

Logarithms are very important in many fields of mathematics and Napier's book *Description of the Wonderful Rule of Logarithms* was quickly adopted by those who had to conduct such calculations on a regular basis. It took Napier an astonishing twenty years to perform the calculations required for the tables of logarithms in the book – that's some dedication.

The logarithm of a number is the number we have to raise 10 by in order to generate that number. For example:

The logarithm of 100 is 2 because $100 = 10^2$

The logarithm of 1,000 is 3 because $1,000 = 10^3$

The shorthand for writing this would be
$\log (1,000) = 3$

We can find logarithms for numbers that are not whole powers of 10 too:

$$\log (25) = 1.39794 \text{ because } 25 = 10^{1.39794}$$

We can also find the logarithm for numbers using any number, not just 10, as a base:

$$\log_5 (25) = 2 \text{ because } 25 = 5^2$$

Natural logarithms (see page 126) are logarithms with a base of e, which is a very special number in mathematics that allows many difficult calculus problems to be solved.

Logarithms today are computed using a calculator or computer, but originally they were worked out either by hand or using Napier's tables. Before desktop calculators became commonplace, people would use logarithms and a **slide rule** (see page 104) to perform difficult calculations.

Napier's bones

Napier also devised a faster and more convenient way of performing multiplication, based on a lattice method that Fibonacci had learned from the Arabs.

Napier's bones was a useful tool that consisted of a set of sticks engraved with numbers, and each stick had a times table written on it, from 2 times the number up to 9 times the number:

If you wanted to multiply 567 by 3, for example, you would collect together the three sticks that match the large number and then highlight the third row:

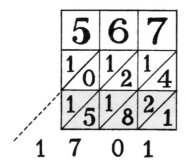

You then add the diagonal rows shown on the bones in order to calculate the answer.

Good Point

Napier was also one of the first people to use the decimal point. Although the Hindu-Arabic numeral system was in common use across Europe by this point, a standard way of writing fractions was yet to be formalized. Because Napier needed a concise way to write them for his log tables, he adopted the Hindu-Arabic decimal fractions we use today.

WILLIAM OUGHTRED
(1574–1660)

An English mathematician and teacher, William Oughtred continued Napier's work on logarithms. He is credited with inventing the slide rule, a calculating device that allowed the user to multiply large numbers together using a ruler marked with **logarithmic scales** (see page 106), which meant the answer to the multiplication could simply be read off the ruler. Slide rules were used by scientists, engineers and mathematicians up until electronic calculators became commonplace in the 1970s.

In action

The slide rule works because of the law of logarithms:

$$\log (a \times b) = \log (a) + \log (b)$$

Let's test this out using some easy numbers:

$$\log (10 \times 1{,}000) = \log (10) + \log (1{,}000)$$

$$= 1 + 3$$

$$= 4$$

From this we know that $10 \times 1{,}000 = 10^4 = 10{,}000$.

Therefore, if you want to multiply two numbers together you can 'take the log' of each number, add them together and then raise 10 by this number. Before the slide rule, you would have had to look up this answer in a big, expensive book of tables – so the slide rule was a remarkably convenient development for mathematicians and scientists.

Jumping ahead

Rather than spacing out successive markings along an axis by adding on the same value each time, a logarithmic scale spaces them out by a multiple (normally of 10) each time. For example, a graph with a normal **linear scale** would go 0, 1, 2, 3; a graph with a logarithmic scale would go 0, 1, 10, 100. The graphs of $y=10^x$ shown below are with a linear and logarithmic scale respectively.

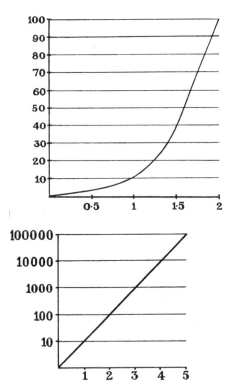

RENÉ DESCARTES
(1596–1650)

Born in France, René Descartes was an important philosopher, perhaps best known for coining the statement, 'I think, therefore I am'.

A different equation

Descartes was a sickly child and was allowed to sleep in every morning to recuperate, which became a lifelong habit. Apparently, during one such lie-in, Descartes was watching a fly walk across the ceiling and wondered how he might be able to describe accurately the position of the fly at any given time. He realized that if he mapped out the ceiling with a square grid he could use what we now call **coordinates** to record exactly where the fly was positioned.

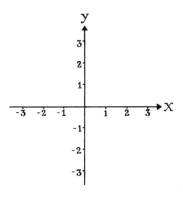

These **Cartesian graphs** proved to be incredibly useful in mathematics because they linked the fields of geometry and algebra. Equations could now be drawn on a set of numbered axes, which allowed them to be investigated more easily by sight, rather than solving them algebraically.

Cartesian geometry encouraged mathematicians to think about the graphical properties of equations, such as whether or not the lines of equations are parallel. To find out whether lines *are* parallel you need to work out their **gradient**. If two lines have the same gradient they must be parallel. Only the simplest linear equations are a straight line, which makes their gradients straightforward to work out. Curves, however, have a changing gradient, but thanks to Descartes' breakthrough the way was paved for Isaac Newton and Gottfried Wilhelm Leibniz to discover calculus (see page 115).

The Proof of the Matter

Descartes' philosophy also assumed a mathematical bent because he believed the universe was set out according to mathematical rules. For example, in his book *Meditations on First Philosophy*, Descartes discards any beliefs he holds that are unproven, and then builds his philosophy of the way in which reality works without recourse to anything other than proven fact. This sceptical viewpoint is one of the ideas that underpins modern science and makes mathematics the scientist's most powerful and productive tool.

PIERRE DE FERMAT
(1601–65)

A French lawyer, Pierre de Fermat spent his spare time pursuing his love of mathematics. Although he did not publish any of his ideas when he was alive, he did share them in letter form with his mathematical contemporaries. Frustratingly, however, Fermat seldom found it necessary to provide proof for his work.

Fermat worked in many areas of mathematics, but the area for which he is famed is **number theory**: the study of **integers** (whole numbers) and the attempt to find integer solutions to equations.

Fermat is famous for his 'last theorem', which he wrote in his copy of Diophantus' *Arithmetica*, and which was discovered by his son after his death (see page 66). Fermat was also interested in **perfect numbers** (see box on page 110) and primes.

In their prime

Fermat devised a method of testing whether or not a number is prime that relies on an algebraic trick known as the **difference of two squares**. This says that:

$$a^2 - b^2 = (a + b)(a - b)$$

Not So Perfect

A perfect number is a number whose factors (not including the number itself) add up to make the number.

For example, 6 is a perfect number (the first one, in fact) because the factors of 6 are 1, 2 and 3 and 1 + 2 + 3 = 6

Perfect numbers are rare – the next one is 28, followed by 496 and then 8,128. The fifth perfect number is 33,550,336.

Numbers whose factors add up to less than the number are called **deficient numbers**. For example, 8's factors are 1, 2 and 4, which have a sum of 7, so 8 is deficient.

Numbers whose factors add up to more than the number are called **abundant numbers**. For example, the sum of 12's factors are 1 + 2 + 3 + 4 + 6 = 16.

In the case of some abundant numbers, no combination of its factors will make up the number. These numbers are called **weird numbers**. For example, 24 is not a weird number because we can add its factors together (2, 4, 6 and 12) to make 24. The first weird number is 70 – we cannot add any combination of its factors 1, 2, 5, 7, 10, 14 and 35 to get a total of 70.

For example, if a = 8 and b = 5:

$$8^2 - 5^2 = (8 + 5)(8 - 5)$$

$$64 - 25 = 13 \times 3$$

$$39 = 39 \checkmark$$

Fermat needed to test odd numbers (because 2 is the only even prime) to see if they were prime. He made the number he was testing, n, equal to the difference of two squares:

$$n = a^2 - b^2$$

Which means that:

$$n = (a + b)(a - b)$$

This shows that n is two numbers multiplied together, in which case n cannot be a prime number unless $(a + b) = n$ and $(a - b) = 1$.

Fermat took the first statement and rearranged it:

$$a^2 - n = b^2$$

This meant that he could pick a starting value for a, square it and subtract n and see if he was left with a perfect square which is easily identifiable. If b^2 was not a perfect square he would increase his starting value by one and try again until he either found numbers for a and b that worked or got to a point where a × b was larger than n.

Blaise Pascal (1623–62)

Educated by his father, Blaise Pascal was a French prodigy who worked in the fields of mathematics, physics and religion. His precocious talent saw his first mathematical paper published at the tender age of sixteen.

Speeding things up

Pascal's father was a tax collector during a time of war in Europe, which made his job a somewhat onerous endeavour. Pascal sought to help his father by developing the first mechanical calculator – a machine known as a 'Pascaline', designed to add and subtract numbers. After he'd created a number of prototypes, Pascal's finished product comprised a box with a series of numbered dials on its front and with a digit displayed above each dial. Numbers to be added were 'dialled' into the machine and the result would be displayed.

Unfortunately, the Pascaline was very expensive to make and was therefore seen as more of a deluxe executive toy than a useful mathematical device. But Pascal's contribution to mathematics should not be underestimated – he paved the way for Leibniz and others to develop more effective mechanical calculators and, eventually, modern computing.

In all probability

Pascal was also interested in games of chance and gambling. His work with his acquaintance Pierre de Fermat (see page 109) led to the field of mathematics we now call **probability**. In probability, we talk about an event (e.g. rolling a die) having a certain number of outcomes (rolling a 1, 2, 3, 4, 5 or 6 has six outcomes). Each outcome has a probability – for our die, each outcome is equally likely – which is expressed as a fraction (⅙), and the sum of the probabilities of all the events must add up to one. Probability is part of the branch of mathematics called **statistics**, which has a wide variety of applications in science and economics.

Under pressure

As a scientist, Pascal was fascinated with the notion of a vacuum. At the time, many scientists conformed to the view expounded by Aristotle: vacuums cannot exist; you cannot have emptiness because 'nature abhors a vacuum'. However, Pascal noticed that if you place a glass beaker upside down in liquid (he used mercury) and then pull it out, there is a small gap at the top of the up-ended beaker that somehow holds up the column of liquid below it. He reasoned that this could only be a vacuum and that it must provide some sort of suction force to hold up the liquid.

Pascal went on to conduct more experiments on pressure within liquids and, as a result, the unit of pressure is called the Pascal (Pa) in his honour.

Absolute Proof

In 1654 Pascal had a profound religious experience and it changed the course of his life. He subsequently devoted himself to an ascetic existence and focused on writing theological commentaries. He used his knowledge of probability to expound a reason for assuming God exists, now known as **Pascal's wager**:

You cannot tell whether God does or does not exist through logic.

If you believe that he exists and he does not, you lose nothing.

If you believe that he does not exist and he does, you lose an eternal afterlife.

Therefore, there is nothing to lose and possibly infinite reward to be had from believing in God and nothing to gain from not believing in him.

So, on balance, you may as well believe God exists.

ISAAC NEWTON (1642–1727)

One of the greatest scientists of his era, Isaac Newton hailed from Lincolnshire and wrote one of the most important books ever to be written: *The Mathematical Principles of Natural Philosophy*, often known by a shortened version of its Latin name *Principia Mathematica*. In the book Newton essentially rewrites the laws of physics that govern the way objects move and react to forces exerted on them. With his laws, Newton was able to explain the motion of the planets and prove conclusively that the sun sits at the centre of the solar system.

Scientifically, Newton's insight was to recognize gravity as a force that is caused by the earth's mass, and his ability to understand intuitively how objects would behave when the earth's gravity was not present. Newton's critics saw gravity, which acts invisibly and at a distance, as some kind of demonic force and that Newton, an alchemist, was obviously in league with such forces. However, Newton's Law of Universal Gravitation and his Equations of Motion were perfectly sufficient to allow us to send men to the moon 300 years later.

Change afoot

Mathematically, Newton's greatest achievement is **calculus**, which was also developed independently by Gottfried Leibniz at approximately the same time. Calculus is a tool used today in a range of different disciplines to describe and predict change. Building on the work of Descartes (see page 107), calculus can be split into two main branches.

1. **Differentiation** involves finding the gradient of the line of an equation. Straight lines have a constant gradient that can be easily measured on a graph.

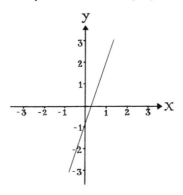

We can see that for every square the line moves to the right (the positive direction), the line goes up two squares. The gradient of the line, therefore, is 2.

However, not all equations give straight lines. Any equation with x^2, x^3 or higher powers produces a curved line:

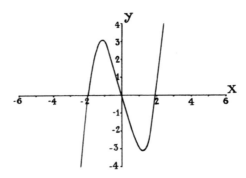

The gradient of this line constantly changes. However, the gradient of the **tangent** – the line that meets the curve at a point – is the same as the gradient of the line at that point:

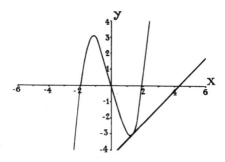

Differentiation lets us find a formula for the gradient of the line at *any* point so we no longer need to draw the tangent, which eliminates an area prone to error.

If we have the formula for the gradient at any point, we can find the places where the gradient is zero. These are called the **turning points** of the equation and finding them can be very helpful. Many problems in mathematics, banking and business involve finding the maximum or minimum values of an equation – differentiation lets us find these points. Scientists have also found that many phenomena are governed by **differential equations**. For example, Newton's Second Law, force = mass x acceleration, is derived by differentiating momentum.

2. The other branch of calculus is **integration**, which is concerned with finding the area between a curve and the x axis:

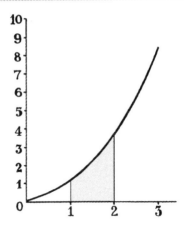

Again, drawing the graph might enable us to estimate the area under the curve, and there are various numerical methods that allow us to calculate an approximation of the area. One method would be to divide the area into thin rectangular strips and add each area together:

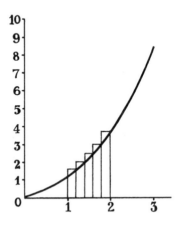

As you increase the number of rectangles you get closer to the actual value. Newton and Leibniz took this one stage further.

They imagined the rectangles became infinitely thin, in which case you get the true value of the area.

You can use this method if you want to calculate the precise area of the shape under a curve. As with differentiation, there are many scientific laws that rely on integration.

Back to the Beginning

It turns out that integration and differentiation are the **inverse** of each other, which means that integrating an equation and then differentiating it again takes you back to the original equation. This is known as the **Fundamental Theorem of Calculus**. Mathematicians use it to help them perform calculus on more difficult equations.

GOTTFRIED LEIBNIZ (1646–1716)

A mathematician and philosopher from Saxony, in the present-day state of Germany, Gottfried Leibniz was the son of a philosophy professor who died when Leibniz was just six years old. Leibniz inherited his father's extensive library, through which he gained much knowledge, after first having taught himself Latin so he could read the books. Leibniz began his working life as a lawyer and diplomat, and, while on secondment in Paris, he met a Dutch astronomer called Christiaan Huygens, who assisted him in his learning of science and mathematics.

Things turn ugly

Leibniz is important for several reasons, although, perhaps unfortunately, he is mainly remembered for the bitter dispute he had with Isaac Newton over the invention of calculus. Newton was based in Cambridge and Leibniz in Paris, where both men devised the concept of calculus independently of each other. Newton began work on calculus as early as 1664 but he failed to publish his findings. That responsibility was left to Leibniz, who published his first paper on the subject in 1684.

The argument centred on whether or not Leibniz had been exposed to Newton's prior work. No evidence exists to prove whether or not Leibniz did have access to Newton's work, and there is no reason to assume that Leibniz could not have come up with calculus independently. However, he died with the matter still unsettled.

Leibniz and Newton developed different notations for calculus and both are used in different areas of mathematics. Leibniz's is perhaps more commonly used.

To evaluate the area shown on the graph you would use Leibniz's notation to write:

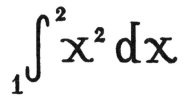

Which is shorthand for: 'integrate x^2 between $x=1$ and $x=2$ with respect to x'.

If you wanted to find the gradient of the line at a point you would need to use differentiation, for which the notation is:

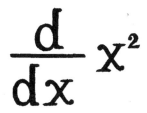

A new dawn

Leibniz was also instrumental in a new field of mathematics that was emerging at the time: computing. In our Hindu-Arabic numeral system, each column in a number is ten times larger than the one on its right – a decimal system. Leibniz was interested in a way of writing numbers in which each column is twice the value of the column on its right – a binary system. The binary system required only the digits 0 and 1, and the columns have values of 1, 2, 4, 8, 16, etc., doubling each time. So the decimal number 13 would be written as 1101:

Column: 8   4   2   1

Digit: 1   1   0   1   because 8 + 4 + 1 = 13

This system seems quite peculiar but it has the advantage of using only two digits, which makes calculations easier. The binary system would later become very important for electronics and computers.

Number Punching

Leibniz also pioneered the 'Stepped Reckoner', one of the first mechanical calculators that could perform multiplication and division. It was a very intricate machine and the complex system of gears could be unreliable, but as manufacturing technology improved over time the Stepped Reckoner went from strength to strength, and Leibniz's ideas were used for hundreds of years, well into the twentieth century.

JOHANN BERNOULLI (1667–1748) AND JACOB BERNOULLI (1654–1705)

The Bernoulli brothers were Swiss mathematicians. Although they both pursued alternative professions – Johann was trained in medicine and Jacob as a minister – both siblings loved mathematics, especially the calculus of Leibniz. The Bernoulli brothers were able to push on with the fledgling field of calculus and became proficient in its use, turning it from an intellectual and political curiosity into a useful mathematical tool. They were also fiercely competitive with each other, which spurred on their discoveries even more.

Arch-rivalry

One example of the Bernoulli brothers' rivalry stemmed from the problem of the **catenary curve**: the shape produced by a rope or chain when it hangs from both ends. A mathematical equation of this shape had eluded mathematicians up to this point. Jacob proposed the problem in 1691; Johann, with some assistance from Leibniz and the Dutch mathematician Christiaan Huygens, then went on to solve it. Catenary curves have important applications in bridge building and in architecture because arches that follow an upside-down catenary curve are the strongest.

In everyone's interest

Jacob also discovered something interesting when he looked at a problem involving **compound interest**. Jacob noticed that if you had £100 in a bank account that paid 10% interest per annum, the way the interest is paid throughout the year affects the total money you will have at the end of the year. Compound interest payments add to the principal sum of money in a bank account, which increases the interest you earn year after year:

Interest	Interest Calculation	Total
10% paid at end of year	100×1.1	£110
5% every six months	$(100 \times 1.05) \times 1.05$	£110.25
2.5% every 3 months	$(((100 \times 1.025) \times 1.025) \times 1.025) \times 1.205)$	£110.38
Daily interest		£110.51

Admittedly, the changes are not making a vast difference to your balance, a fact that most banks rely on. Of greater significance was Jacob's investigation into instances when interest is paid continuously, in tiny amounts, over the entire year. He discovered that if your interest rate is x (e.g. x=0.045 for a rate of 4.5%) at the end of a year you would have 2.718281^x times what you started with. I have rounded the 2.718281 – it is in fact an irrational number, like π, that goes on for ever without repeating.

Napier made reference to this number in his work with logarithms, and it also plays a very important role in calculus, as exemplified by our next mathematician.

LEONHARD EULER (1707–83)

Euler (rhymes with 'boiler' rather than 'ruler') was a Swiss mathematician who had originally intended on becoming a priest. However, while at university he met Johann Bernoulli, who recognized Euler's extraordinary mathematical talent and managed to persuade Euler senior to allow his son to transfer to studying maths.

An increase in power

Euler's contributions to mathematics and science were far-reaching. The number 2.718281..., discovered by Jacob Bernoulli in relation to his compound interest problem, also turned up in Euler's work on calculus.

When you integrate to find the area under a graph, you need to increase the power of x by one. For example, if your graph is $y=x^2$, the integral is $\frac{1}{3} x^3$ – the power of x has gone up by one. If you're faced with something slightly more tricky, let's say $y = \frac{1}{x^4}$, there is a handy rule of powers that can help you:

$$1/x^n = x^{-n}$$

So $y=1/x^4$ becomes $y=x^{-4}$, and when you increase the power by one your answer will be something to do with x^{-3}, which is $1/x^3$.

But what happens if your graph is $y = 1/x$?

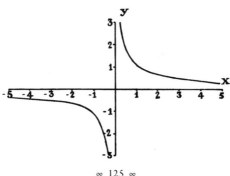

This is the same as x^1, but if you increase the power by one you get x^0. Anything to the power of zero is 1, implying that, no matter which section of the graph you look at, the area will be the same. This doesn't make sense!

Well, it turns out, through a complicated system of algebra developed by Euler, that the area is equal to the natural logarithm of x. A natural logarithm is similar to a normal logarithm, but its base is the number 2.7818281... There is a sizeable family of equations that can only be integrated or differentiated using natural logs, and the 2.7818281 was known as 'e' for Euler's number.

Returning to the area under the graph, if you wanted to know the area between x=1 and x=4, you would need to work out:

$$\text{area} = \log_e 4 - \log_e 1$$

As '$\log_e$' turns up so often in calculus, it is denoted by ln and you will find this button on all good scientific calculators.

$$\text{area} = \ln 4 - \ln 1 = 1.386 \text{ (to three decimal places)}$$

A bridge too far

Euler's work on 'The Seven Bridges of Königsberg' contributed to methods of simplifying maps. Königsberg was the old Prussian name for the city of Kaliningrad in that strange bit of Russia that sits between Poland and Lithuania. The city is centred on an island, which straddles a river. Seven bridges connect the two sides of the island at various locations:

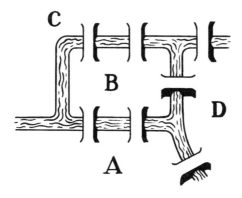

A popular Sunday afternoon pastime for the residents of Königsberg was to attempt to walk over all seven bridges and return to their starting place without having to use the same bridge twice. Not one person ever managed it, but Euler was the first to tackle the problem mathematically. He redrew the map as a network:

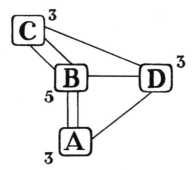

Euler counted the number of routes into and out of each node or intersection on his network. He then reasoned that you would have to walk into and out of each node as you made your way around the map, so you needed each node to have an even

In My Imagination

Euler was also interested in what are known as complex numbers. These are numbers that are made up of two parts, one real (i.e. any number between plus and minus infinity) and one imaginary.

Diophantus had the first inklings of imaginary numbers (see page 65), but it wasn't until the sixteenth century and the arrival of two Italian mathematicians, Niccolò Fontana Tartaglia and Gerolamo Cardano, that the study of imaginary numbers really took off. Tartaglia and Cardano discovered that some equations only generate an answer if you are prepared to allow negative numbers to have square roots – which can't happen with real numbers, because a negative multiplied by a negative gives a positive.

Descartes coined the term 'imaginary' – even though such numbers don't exist, you can permit them to exist in your imagination in order to find answers to previously unsolvable equations.

The letter i is used to denote the square root of minus 1: $\sqrt{-1}$

This allows you to reference the square root of any negative number:

$\sqrt{-49} = \sqrt{(49 \times -1)} = \sqrt{49} \times \sqrt{-1} = 7i$

Despite these numbers being imaginary, they have many practical applications, especially in electronics and electrical engineering.

number of routes. All the nodes in Königsberg have an odd number, so it is impossible to complete the challenge. Networks where all the nodes have even numbers are known as **eulerian**. A **semi-eulerian** network is one that possesses two nodes with an odd number; if you start at one odd node and finish at the other you can complete the network without repeating yourself, as with the following famous example:

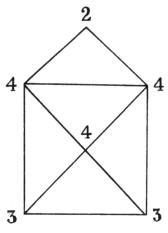

This way of simplifying and thinking about maps and routes had implications for cartography. It's also an important part of **decision mathematics**, the branch of mathematics that many businesses rely on for planning delivery routes and other logistical operations.

The many faces of geometry

Euler worked in three-dimensional geometry too. He discovered there is a relationship between the number of corners (or vertices), edges and faces of a polyhedron (a three-dimensional shape with flat faces, like a cube or a pyramid):

$$\text{vertices} - \text{edges} + \text{faces} = 2$$

Mathematical Perfection

Euler devised an equation now known as **Euler's identity** (an identity is an equation that is always true no matter what value you use for the unknown), which is said to be the most beautiful and elegant mathematical equation. It relates to complex numbers, but unfortunately its meaning is beyond the scope of this book. Here is the equation in all its glory:

$$e^{i\pi} + 1 = 0$$

Its subjective mathematical beauty arises from the fact that it uses five of the most important numbers in mathematics: e, i, π, 1 and 0.

A cube has 8 vertices, 12 edges and 6 faces: $8 - 12 + 6 = 2$

A tetrahedron (triangular-based pyramid) has 4 vertices, 6 edges and 4 faces: $4 - 6 + 4 = 2$

A dodecahedron (12-sided polyhedron) has 20 vertices, 30 edges and 12 faces: $20 - 30 + 12 = 2$

A truncated icosahedron (the combination of hexagons and pentagons used to make a football) has 60 vertices, 90 edges and 32 faces: $60 - 90 + 32 = 2$

Euler also teamed up with Daniel Bernoulli (1700–82, son of Johann, nephew of Jacob) to work in applied mathematics. They considered the forces acting on beams in buildings and how the forces would make the beams bend – a very useful tool in engineering applications.

True or False?

Mathematician Christian Goldbach (1690–1764), in a letter to Euler about the nature of prime numbers, wrote what has become known as **Goldbach's conjecture**:

'Every even whole number greater than 2 can be written as the sum of two prime numbers.'

For example, 10 can be made up from 5 + 5 and 28 is 11 + 17.

In mathematics, ideas are split into three categories:

1. **Propositions** are statements that may or may not be true. Euclid proposed many in his *Elements* that he then showed to be true.

2. When a proposition has been shown to be true in all possible cases, it is said to be a **theorem**, like Pythagoras' – it works for *all* right-angled triangles.

3. A **conjecture** is a proposition that holds the middle ground – mathematicians believe it to be true but have not yet been able to prove it is always true.

Although Goldbach's conjecture has been checked as far as 4,000,000,000,000,000,000 without finding a counter-example, it is still only a conjecture rather than a theorem. Very picky, these mathematicians!

CARL GAUSS (1777–1855)

Carl Gauss was born into a poor family in Germany in 1777 but it soon became apparent that he possessed an extraordinary intellect and a special ability in mathematics in particular.

According to legend, when Gauss was at school he continually annoyed his maths teacher by completing his work far faster than the rest of his class. Exasperated, Gauss' teacher finally told him to add together all the numbers from 1 to 100, thinking that might give him some peace.

Gauss immediately stated the correct answer: 5,050.

Gauss was not a lightning calculator; he had instantaneously seen a short cut. If you repeat the series, but backward, it can be seen that all the terms add up to 101:

$$1 \; + \; 2 \; + \; 3 \; + \; ... \; + \; 98 \; + \; 99 \; + \; 100$$
$$100 \; + \; 99 \; + \; 98 \; + \; ... \; + \; 3 \; + \; 2 \; + \; 1$$
$$101 \; + \; 101 \; + \; 101 \; + \; ... \; + \; 101 \; + \; 101 \; + \; 101$$

Gauss then quickly worked out that 100 terms of 101 gives 10,100, but as this is twice the sum we actually want, he halved it to give the answer 5,050.

Downing Tools

When Gauss was at university he was interested in the classical geometry of the ancient Greeks, but new developments in mathematics meant it was now possible to prove geometric theories using algebra, rather than graphically. Gauss proved that it was possible to draw a regular polygon (all sides of equal length and all interior angles of equal size) with 5, 17 or 257 sides using only a pair of compasses and a straight edge.

Back to the beginning

Gauss furthered Euler's initial work in a branch of number theory called **modular arithmetic**, in which numbers are allowed up to a certain value, after which they wrap-around and start again. The twenty-four-hour clock is an example of modular arithmetic – after 23:59 we start again at 00:00.

Much like normal arithmetic, in modular arithmetic you need to define what your highest number can be. In normal arithmetic we work in tens, but our highest digit is one less than this: 9. If we are working in modulo 8 we can only use the digits from 0 to 7. This means that 8 would be 0 in modulo 8 because we start again from zero after we reach 7. Likewise, 15 would be 7 in modulo 8 because 15 = 8 + 7 but the 8 counts as 0. Mathematicians would write this as:

$$15 \equiv 7 \ (\text{mod } 8)$$

In certain circumstances, if you divide two numbers you may be more interested in the **remainder** (what's left over) than the **quotient** (the answer to the division). This is where modular arithmetic can be useful, because a number's value in a particular modulo is the same as the remainder if you were dividing it. For example:

$$75 \div 8 = 9 \text{ remainder } 3$$

$$75 \equiv 3 \ (\text{mod } 8)$$

If you wanted to check whether a number was prime, you could see whether the number was ever equal to zero in successive modulos, which is something that computers are good at.

A magnificent spread

Gauss' work naturally moved into prime numbers, which remain one of the greatest mysteries in mathematics. Gauss made a conjecture, now called the **prime number theorem** (it is a theorem because it has since been proven, see page 132), about the way in which prime numbers are spread out. Although we do not have a formula for making prime numbers, Gauss noticed that the higher up through the numbers you go, on average the more spread out the prime numbers become. He wrote:

$$\text{number of primes less than } x \approx x \ / \ \ln x$$

The symbol ≈ means 'is roughly equal to' and the symbol ln means 'natural logarithm'. Therefore:

> number of primes less than 1,000 ≈ 1,000 / ln 1,000 ≈ 145
>
> number of primes less than 10,000 ≈ 10,000 / ln 10,000 ≈ 1,086

This shows that, although we made x ten times larger, there are fewer than eight times as many primes. This trend continues as we make x bigger, so primes become fewer the higher we count.

Uneven distribution

Gauss also made an important contribution to statistics by being the first person to introduce the **normal distribution**. This bell curve applies to all manner of real-world situations such as animals' heights and weights, marks in examinations, measurements made in scientific experiments and so on.

If you measured the height of every thirteen-year-old boy in the country, you could work out the average or **mean** height (worked out by adding up all of the data and dividing by how many data there were). You could then look at the percentage of the boys in a certain height bracket and you would find that most of them were within a certain distance from this mean. As you move away from the mean, either higher or lower, you find that there are fewer and fewer boys. Thinking in terms of percentages like this is the same as thinking in terms of probabilities, and so the normal distribution is said to be a **probability density function**:

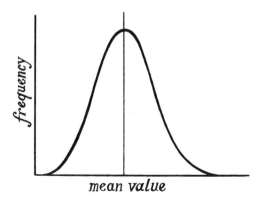

mean value

The shape of the graph shows what we know to be true. Think about your own friends. Unless you hang out with professional basketball players, most of your friends are clustered around an average or normal height for their gender, and you probably know far fewer very tall and very short people.

The idea of IQ (intelligence quotient) is an example of a score that has been standardized using the normal distribution. An IQ score of 100 is the mean, and a 15-mark interval is called the **standard deviation**, which is a measure of how spread out the marks are. As a result of the equation of the line used for the normal distribution, it turns out that over 68% of scores are within one standard deviation of the mean, so nearly 70% of people will have an IQ of between 85 and 115. Over 95% of people are within two standard deviations, scores of 70 to 130. Over 99.7% of people are within three standard deviations, scores of 55 to 145. Mensa, a society for people with high IQs, has an entrance test intended to select people who have an IQ higher than 98% of the population, which corresponds to an IQ of just under 131.

A Question of Identity

An **equation** (such as $y + 3 = 10$) is something we can try to solve in order to find out if there are any values that satisfy the equation. **Linear equations**, where the unknown has an index of 1 (i.e. it is not squared or cubed), have only one answer. Equations that incorporate squares, or cubes, or higher can have more than one answer, but equally may have no answer. For example, there is no real number x that works in $x^2 = -6$.

In **formulae** (such as Einstein's $E = mc^2$ or average speed = distance ÷ time) we can substitute values for the letters in order to solve the equation. For example, if you travelled 200 kilometres in 4 hours, you would get the following:

Average speed = 200 ÷ 4 = 50 km / hour

An **identity** is something that is always true for any value of the unknowns. Gauss invented the triple-bar symbol, $\equiv$, to show this. For example:

$(y + 2)(y - 3) \equiv y^2 - y - 6$

This is an identity because it works for any value of y. Say I make $y = 7$:

$(7 + 2)(7 - 3) = 7^2 - 7 - 6$
$9 \times 4 = 49 - 13$
$36 = 36 ✓$

The Digital Age

Although they didn't become commonplace in homes until the 1980s, one can't now imagine a life without computers. The history of the computer is intimately linked with the story of mathematics. We have seen already that pioneers like Leibniz and Pascal blazed (sorry...) a trail with mechanical calculators, but the need for faster and more flexible machines sparked off the digital revolution.

CHARLES BABBAGE (1791–1871)

The English mathematician Charles Babbage was the son of a banker, and is best known as the 'Father of the Computer' because of his work in mechanical computing. Babbage read mathematics at Cambridge University but found his studies unfulfilling, which led him to found the Analytical Society in 1812 along with a group of fellow students. Apparently,

Babbage's moment of inspiration came when he was one day poring over a group of logarithm tables that were known to contain many errors. Working out the entries for log tables is a very tricky process and inevitably human 'computers' – as they called people who calculated sums in those days – made mistakes. Babbage maintained that the tedious calculations, which require little thought, only accuracy, could be performed by an elaborate calculating machine.

Machine takes over

Babbage secured government funding and produced plans for his **Difference Engine**. However, such was the complexity and size of the machine that it was never built in his lifetime. (A Difference Engine was built in the 1980s and today resides in London's Science Museum. At just over 2 metres high and 3.5 metres long it's a fairly sizeable machine.)

As interest in the Difference Engine project began to wane, Babbage started work on the Analytical Engine using the knowledge he had gleaned while designing his first computer. This machine was intended to be much more like a computer as we know it. It could be programmed to perform particular combinations of mathematical functions. One set of punched cards would be used to programme the engine, another set would introduce data to the engine and then the machine itself could punch blank cards with the results, effectively saving them in its memory for future use.

Babbage was still working on the Analytical Engine when he died. Because of the immense cost and complexity involved, a working model has never been built, although, at the time of writing, British programmer and mathematician John Graham-Cumming is leading a project to build one for the first time.

ADA LOVELACE (1815–52)

One of the few notable pre-twentieth-century female mathematicians, Ada Lovelace was the daughter of the famous poet and rake Lord Byron and Anne Isabella Milbanke, to whom Byron was briefly married. Milbanke was convinced Byron's excesses had been the result of a kind of insanity and she sought to shield her daughter from a similar fate by encouraging her to pursue a persistent education, especially in the logical discipline of mathematics.

Get with the programme

Despite being unable to enter university, Lovelace was privately tutored and she continued with mathematics throughout her life. She encountered Charles Babbage, who asked her to translate into English an article on his Analytical Engine that had been written by an Italian mathematician. Lovelace duly did so, adding extensive notes on the machine, including a set of instructions that would have had the machine produce the **Bernoulli numbers**, a sequence of numbers named in honour of Jacob Bernoulli. For this, Lovelace is considered to be the first ever computer programmer. Sadly she died from cancer at the age of thirty-six.

Ockham's Razor

When Ada Lovelace married William King in 1835 she moved to her husband's estate in Ockham, Surrey, which is believed to have been the birthplace of a monk known as William of Ockham in the late thirteenth century. William of Ockham was a natural philosopher who first coined the principle known as **Ockham's razor**: the simplest explanation for a phenomenon more often than not turns out to be true. This principle has been adopted in all scientific fields – when researchers look to explain what they have observed, they try to use existing theories and laws rather than fabricate new ones to fit what they have seen.

GEORGE BOOLE (1816–64)

An Englishman who became a mathematics professor in Ireland, George Boole published major works on **differential equations** (equations that involve derivatives of a function), but he is most remembered for his work on logic.

Boole sought to set up a system in which logical statements could be defined mathematically and then used to perform mathematical operations on the statements; the results would be generated without the need to think through the problem intellectually. Boole's system aimed to take a raft of logical propositions and see how they combined together with the aid of maths rather than philosophy.

A logical step

In order to set up the system, Boole developed what became known as **Boolean algebra**: letters defined either as logical statements or as groups of things. The letters can have a value of 1, meaning 'true', and 0, meaning 'false'.

So imagine you are considering dogs as a group of things, and you let x represent shaggy dogs and y represent yappy dogs. You can then make a table for the dogs using values for x and y:

x	y	Notes
0	0	Neither shaggy nor yappy
0	1	Not shaggy but yappy
1	0	Shaggy but not yappy
1	1	Shaggy and yappy

Boole then defined three simple mathematical operations we could conduct with the results – AND, OR and NOT. AND is defined as the multiple of the two values. As the table below shows, if x and y are 'true', we expect the answer 'true' or 1, and everything else to be 0 or 'false':

x	y	x × y
0	0	0
0	1	0
1	0	0
1	1	1

This seems rather self-evident – a dog can only be shaggy AND yappy if the dog is shaggy and yappy, but the example is useful because it shows you can come to the same conclusion using very simple arithmetic.

There are a host of other Boolean operations that can be reduced to arithmetic.

In the 1800s Boole's work had implications for mathematical logic and set theory, but it was not until the twentieth century that a more practical use was found. An electronics researcher called Claude Shannon discovered that he could use Boolean operations in electrical circuits – he could generate a simple electrical circuit to take logical steps and therefore make decisions based on them. As Boole's work uses only values of 0 or 1, 'true' or 'false', or 'on' or 'off', it is known as a **digital** method. In essence, Boole paved the way for the first electronic digital computers.

Setting things out

Set theory is the branch of mathematics concerned with placing objects into groups called sets. Sets can be defined as containing:

1. numbers (e.g. all odd numbers less than 100)

2. objects (e.g. the set of different types of dogs)

3. ideas (e.g. the set of problems that can be solved by a computer)

Boolean algebra works in set theory, like the example of the shaggy and yappy dogs on page 143.

John Venn (1834–1923) was a British priest and logician who, besides designing and building a cricket-ball bowling machine, is best known for **Venn diagrams**, which help to show set theory in a visual way. To use our earlier example, if you are considering all types of dog, then the rectangular box should be receptive to all types of dog. I introduced two sets: x, the set of shaggy dogs, and y, the set of yappy dogs. We saw that some shaggy dogs are also yappy, so those two sets need to overlap.

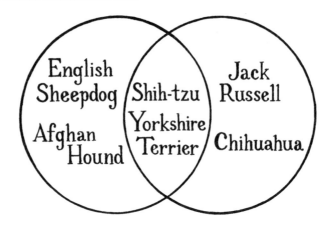

The overlap in the middle represents x AND y. x OR y would be any dog within the two circles. NOT x would be any unshaggy dog.

To bring things back into more mathematical territory, the German mathematician Georg Cantor (1845–1918) was very interested in infinite sets, which, as the name suggests, are sets containing an infinite number of things.

Consider the set of positive whole numbers (1, 2, 3...) which mathematicians call the natural numbers. This is an infinite set because we can keep on counting for ever. Then consider the set of numbers in which the numbers can be anything at all, including fractions, negative numbers and those irrational numbers like π, e and φ that we saw earlier (mathematicians call this continuous list of numbers real numbers). This is also an infinite set, and it contains more numbers than the set containing the natural numbers. By comparing the set of natural numbers, which is infinite, with the set of real numbers, which is also infinite, Cantor was able to show that, although both were infinite, they were not the same size. Therefore we have the idea that there are different infinities for different infinite sets. This sort of business is the reason many mathematicians consider infinity to be a concept, rather than a number.

Cantor said that the natural numbers were **countably infinite** because we make progress towards infinity as we count. The real numbers are **uncountably infinite** because, no matter where you start counting from, you will not make much progress as you add on an infinitesimal amount each time.

Cantor developed the **continuum hypothesis**, which states:

> There is no set that has more members than the
> set of integers, but fewer members than
> the set of real numbers.

So far, it has been proved mathematically that this hypothesis cannot be proved or disproved within the current limits of standard set theory, and it remains one of the greatest unsolved problems in mathematics.

ALAN TURING
(1912–54)

Turing was a brilliant British mathematician and scientist who is well known for the instrumental role he played in the Allied effort to break the German Enigma cipher, a coding machine used by the Nazis during the Second World War.

Memory machine

Prior to the First World War, Turing was a mathematics fellow at Cambridge University. He worked on the Entscheidungsproblem, a challenge set by the German mathematician David Hilbert (see page 183) that asked whether it is possible to turn any problem in mathematics into an algorithm that will produce a 'true' or 'false' answer that doesn't require a proof. Turing was able to show that it is not possible by introducing the concept of an idealized computer called a **Turing machine**.

A Turing machine is a computer that has an infinite memory that can be fed an infinite amount of data. The machine can then modify the data according to a simple set of mathematical rules to give an output. Turing showed that it was not possible to tell whether a Turing machine would reach an answer to the Entscheidungsproblem, or whether the machine would carry on working out the problem for ever.

The theoretical Turing machine was very important in establishing computer science. Turing's research into computing took him to Princeton University in the United States, where he worked towards a PhD in mathematics. Here Turing built one of the first electrical computers using Boolean logic (see page 143), a crucial development towards what came next.

Unpicking the code

Turing was part of the Government Code and Cipher school based at Bletchley Park in Milton Keynes, UK, during the Second World War, where he was set to work on breaking the Enigma cipher. While there Turing developed something called the **Bombe**, an electro-mechanical machine that could work its way through the huge number of possible settings of the Enigma machine far faster than a human cryptanalyst could. The Bombe relied on the fact that the Enigma would never encrypt a letter as itself – if you typed the letter 'q' into the Enigma, 'q' wouldn't be given as an output. The Bombe would try each setting, and when the output letter matched the input letter that setting would be eliminated and the next one tried. The second trick was to use a **crib**: making an educated guess as to what the first words of an intercepted message might be.

The work of Turing and others at Bletchley Park was hugely important to the Allies and certainly shortened the war. However, the work was top secret and Turing could therefore receive little recognition for his efforts.

Man vs machine

After the war Turing continued his work on electronic computers, turning his hand to artificial intelligence – whether or not a sufficiently powerful and fast computer could be considered intelligent. During his investigations, Turing devised what has become known as the **Turing test**. The test states that a computer is to be considered intelligent if a human being communicating with it does not notice that the computer is not a human being. Turing's work in computers laid the foundations for the digital revolution we now enjoy (and also take for granted). The computer I'm typing this book on, for example, has its origins in a Turing machine.

Into the Chaos

In 1952 Turing turned his hand to biology. In many living things there is a stage when cells change from being very similar to one another to becoming more differentiated. For example, in a developing embryo a group of identical stem cells transform into cells that go into developing the body's organ system. Turing was able to show that this process, called **morphogenesis**, is underpinned by simple mathematical rules that, nonetheless, can develop very complex animals. This idea was well ahead of its time and many people thought Turing was wasted on such research. However, many years after his death, his work on morphogenesis would be recognized as the one of the first glimpses of **chaos** mathematics.

Unfortunately, in 1952 Turing was convicted of gross indecency for homosexual acts, which were deemed illegal at that time. Turing chose to undertake a course of oestrogen injections over a prison sentence. Horrific side effects to the injections led Turing to become depressed, and in 1954 he was found dead of cyanide poisoning. Turing, the founder of modern computing, the unsung war hero, was dead, most likely by his own hand, at the age of forty-one. We can only ponder what discoveries were denied to us by his tragically premature demise.

MAPPING THINGS OUT

Computers were not immediately adopted by mathematicians because many detractors felt that an elegant proof of a problem should be shown formally, rather than by number-crunching. One of the first conjectures proved with the assistance of computing power was called the **four-colour theorem**.

In the nineteenth century South African mathematician Francis Guthrie (1831–99) was shading in a map of England's counties when he noticed that each county could be defined from its neighbouring county using just four colours. Guthrie posited this as the four-colour conjecture, which, when put more formally, states that in the instance of a flat two-dimensional map, on which any regions that share a border cannot be the same colour, no more than four colours are ever needed. It is important to note that regions meeting across a point do not share a border and therefore can be the same colour, and that all the countries on the map are independent, so they can be any colour required (unlike the part of Russia where old Königsberg is).

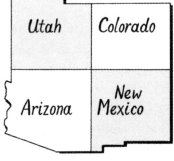

One of the main difficulties with this problem lies in the fact that it is highly visual; there must be an infinite number of maps one could draw to test the theory. However, we can visualize each region as a point using **graph theory** (much like Euler used with the Königsberg problem).

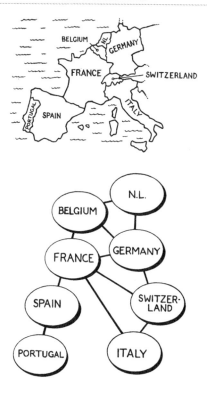

This allowed mathematicians to start tackling the problem in earnest because various different maps could be shown to have effectively the same graph. Serbian mathematician Danilo Blanuša (1903–87) discovered two maps that needed more than three colours. These elusive maps are known as **snarks** and they helped to demonstrate that the minimum number of colours required must be four or more. There are eight known snarks, the last of which was discovered in 1989.

Without a doubt

In the 1970s two researchers in the United States, Kenneth Appel (1932–) and Wolfgang Haken (1928–), then tackled the four-colour conjecture. They discovered that the most complicated part of any map – any area that might need lots of colours, and which might, therefore, disprove the four-colour conjecture – could always be shown as one of 1,936 possible situations. All of these 1,936 situations could be modelled as a graph and, using a computer, Appel and Haken showed that they could all be drawn using just four colours.

So, Appel and Haken showed that any map could be shown to be one of the 1,936 standard ones, so no counter-examples exist.

This was the first theorem proved with the assistance of a computer, but many mathematicians were unimpressed because the exhaustive proof required hundreds of pages of calculations. This meant that checking the proof would take so long it was necessary to trust that the computer's work was correct, which was not an assumption many mathematicians would make at the time.

A PIECE OF THE PIE

π, as every schoolchild knows, is a special number that you get by dividing a circle's **circumference** (perimeter) by its **diameter** (the distance across the middle of the circle). Because all circles are mathematically **similar** (in proportion) you always generate the same value, no matter the size of the circle.

Mathematicians have always been fascinated by π. It is an irrational number (see page 42) and it is a very useful constant in mathematics that occurs not only in questions about circles but also in all kinds of problems in geometry and calculus. Throughout the history of mathematics we have seen people trying to calculate π to increasing levels of accuracy (see box on page 156). My calculator says:

$$\pi = 3.141592654$$

Since the advent of mechanical and electrical computers, the ability to perform the necessary calculations quickly and accurately has improved exponentially. The Chudnovsky brothers from the USA were the first to get π to 1 billion decimal places using a homemade super-computer in the 1980s. The record is currently owned by Japanese mathematician Yasumasa Kanada, who reached 2.7 trillion (2,700,000,000,000) decimal places in 2010.

π Crunching

Here's how the value of π has been calculated across the centuries:

In 1900 BC the Ancient Egyptians calculated a value of 256/81 = 3.1605.

The ancient Babylonians had the handier 25/8 = 3.125.

We have inferred from the Bible an approximation of 3.

We saw that Archimedes placed π between 3.1408 and 3.1429, and he also gave us the common school approximation of 22/7.

In China in the fifth century AD, Zu Chongzhi used 355/113 = 3.1415929.

In c. AD 1400 Indian mathematician Madhava calculated 3.1415926539.

The 100 decimal places mark was reached by English astronomy professor John Machin in 1706.

Swiss mathematician Johann Lambert proved π is irrational in 1768.

Into the unknown

Chaos theory is the branch of mathematics that deals with unpredictable behaviour. Although this field has really opened up since the advent of computers, chaotic equations were first noticed during Newton's time.

One of the main aims of Newton's work was to create a system of equations that explained the motion of the planets. As we know, the planets move around the sun in almost circular orbits. Astronomers had noticed, however, that the orbits would wobble slightly, apparently at random, from time to time.

Leonard Euler (see page 124) developed a concept that became known as the **three-body problem** in an attempt to predict the somewhat irregular orbit of the moon. The reason for this irregularity is because the moon's orbit is influenced by the gravity of both the earth and the sun. The force the sun exerts on the moon varies according to where the moon is in its orbit around the earth, and as a result the moon's orbit fluctuates. Euler was able to work out the governing equations for the situation in which two of the bodies (the earth and the sun) are in a fixed relationship when compared to the third (the moon). French mathematician Henri Poincaré (1854–1912) tried to extend this to a more general solution that could be applied to the whole solar system. He did not succeed, but he was able to show that the orbits would never settle into a regular pattern.

The motion of the planets was not the only phenomenon scientists had observed to be irregular. The mathematics of **turbulence** had long eluded mathematicians. When a gas or a liquid flows, the motion of a particle within it can be explained mathematically under certain circumstances – when the speed of the fluid is relatively low. However, as the speed increases, the motion of a particle, particularly around an obstacle, becomes impossible to predict. This turbulence is highly chaotic. The study of the motion of gases and liquids is called **fluid mechanics** and is important to human beings in all sorts of situations, including transport, electricity generation, and even in understanding the flow of blood around our bodies.

The weather is perhaps the biggest earthly example of turbulence. In the 1960s American meteorologist Edward Lorenz (1917–2008) was modelling the movement of air using a computer when he noticed that, if he changed the initial conditions of his simulations by a negligible amount, as time went by the weather predicted by his simulations would vary widely. This led Lorenz to coin the term 'the butterfly effect': the notion that a minuscule change in an air pattern, much like the change induced by a butterfly flapping its wings, could lead to a hurricane happening elsewhere in the world.

It is this sensitivity to initial conditions that causes unpredictable, chaotic behaviour. Even when we completely understand the motion of the particles involved, we can never gather enough detail in our measurements of their speed or mass or temperature to be able to predict an accurate long-term outcome.

The same applies to Turing's observations in morphogenesis – although the formation of stripes on a zebra is governed by very simple mathematical and chemical rules, immeasurable, tiny differences in each zebra means that they have huge variation in their stripes.

Weather Forecast

The weather is chaotic – we will never be able to predict it accurately for more than a few days in advance, if that, and even then it is possible that the true outcome could be significantly different than predicted. The Great Storm in Britain in 1987 saw the worst winds for hundreds of years, and yet it was not anticipated.

The computer has been key in the study of chaos mathematics because it allows you to take a model of a situation and run it forward, recalculating the many variables at each stage. Without a computer it can be incredibly time-consuming to perform each step, called an **iteration**. Computers are good for iterative processes, which helps in the next branch of mathematics we shall look at: **fractals**.

Under the microscope

Although the term 'fractal' wasn't coined until 1975, mathematicians have been fascinated by them for hundreds of years. A fractal is a geometric figure that has what mathematicians call **self-similarity**: no matter at what scale you look at the image, you still see the same features.

A coastline is a good example of a fractal. Imagine you had a satellite image of a coastline. Without any clues like buildings, trees or boats, you would find it hard to tell whether you were looking at 1,000 miles of coast, or 10 miles of an island. This is because the features – the river inlets, headlands, bays, etc. – look very similar at different scales. The same is true of many features of the natural landscape. Astronauts on the moon found it very hard to gauge the size of boulders – without the haze of atmosphere, they could not tell whether they were looking at a car-sized boulder 100m away or a much larger boulder 500m off.

Everyday Fractals

Many trees and plants have a branching, fractal-like structure. Ferns are especially fractal – the leaves of the plant look exactly like smaller versions of the fronds.

The CGI (computer-generated images) that we enjoy in today's games and films often use fractal-algorithms to make landscapes, foliage, clouds and even skin and hair look realistic.

Famous fractals

Swedish mathematician Helge von Koch (1870–1924) devised one of the first mathematical fractals, known as the **Koch snowflake**, which follows a very simple set of rules: you start with an equilateral triangle, and then with each iteration you add another equilateral triangle to the middle third of each line in the diagram. This builds up a snowflake structure as shown:

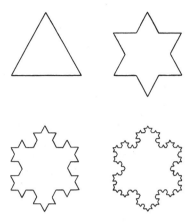

Besides looking pretty, this snowflake has some very interesting properties. It has the **self-similarity property** – if you zoom in on an edge you would not be able to tell how many iterations had been performed. If you continued the iterations ad infinitum, the snowflake would have an infinite perimeter even though the snowflake has a finite area.

Polish mathematician Wacław Sierpiński (1882–1969) invented another straightforward geometrical fractal, again based on a triangular theme. For **Sierpiński's triangle**, you start with a solid equilateral triangle. You then cut out an equilateral triangle from the middle, which gives you three smaller triangles. At each iteration you cut a triangle out of the middle of any remaining triangles to produce the fractal.

Again, we see self-similarity, as each part of the triangle looks like the whole.

It turns out that there are two other unexpected ways to draw Sierpiński's triangle. The first harks back to Blaise Pascal and his triangle. If you shade in all the odd numbers in the triangle, you get Sierpiński's triangle.

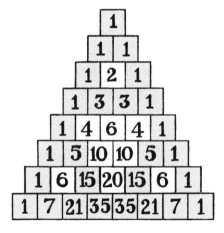

The second way is an example of a **Chaos Game**. You draw the three corners of a triangle and pick a random point inside it. You then, again at random, pick one of the corners of the triangle and move halfway towards it, making a dot at your new position. If you continue this iterative process, you build up Sierpiński's triangle again! It can take a while by hand, but a computer can do this sort of thing very quickly.

The ultimate fractal

There are many other fractals, but the daddy of them all was created by French-American mathematician Benoit Mandelbrot (1924–2010). As a researcher for the computer company IBM, Mandelbrot was initially interested in self-similarity because he felt it was a common feature of many things, including the stock markets and the way the stars are spread out across the universe. Mandelbrot's access to the computers at IBM enabled him to perform many iterations quickly, which allowed his studies to thrive. In 1975 he coined the term 'fractal'.

In the early 1980s he used a computer to investigate for the first time the fractal known as the **Mandelbrot set**. Like so many of the fractals we have looked at, it has a remarkably simple basis:

$$z_{n+1} = z_n^2 + c$$

This iterative equation says that you get the next number in the sequence, z_{n+1}, by taking the current number z_n, squaring it and adding another number, c, to it.

You choose a number, c, and test it using the iteration (starting with $z_0 = 0$). Some values of c cause z to get larger and larger, but for others it gets closer and closer to a stable value or else keeps moving around without ever getting larger than 2.

z and c are both complex numbers, which means they have both a real part and an imaginary part (see page 128). So each value of c has, in effect, two values, which we can plot on a graph as a point. If you plot all the points where c doesn't ever lead to a value larger than 2, you get an image like this:

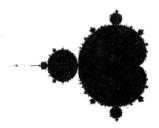

If you shade the unstable points on the graphs to indicate how many iterations it took them to become larger than 2 (the brighter the shade, the more iterations), the famous image emerges.

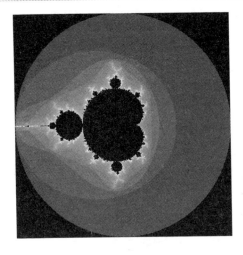

We can see many self-similar features on the Mandelbrot set, but what is perhaps most amazing is the complexity that arises even though the governing equation is so simple. You can literally zoom into the fractal for ever, a mystery that has led to this fractal assuming the mantle, the Thumbprint of God.

As we have seen, the increasing use of computers has also enabled mathematicians to find numeric solutions to equations that cannot be solved using algebra. It is now commonplace for scientists, engineers, designers and inventors to use computer methods to help them in their work. The experiments at the Large Hadron Collider in Europe produce 15 million gigabytes of data each year.

This experiment and many others that require a large amount of number-crunching are now able to exploit the redundant computing power of home computers via the Internet. This **distributed computing** allows the data to be dealt with far more quickly and allows members of the public to be involved in the frontiers of scientific research.

Modern Mathematics

In the last 100 years mathematics has diversified into a vast number of fields, each of which requires years of work in order to excel at it – gone are the days of the generalist mathematician with equal facility across all fields. While a significant proportion of these mathematical developments are beyond the scope of this book – and, if I'm being perfectly honest, beyond the scope of this author, too – there are some interesting areas where, initially at least, we can appreciate the mathematics and the mathematicians that contributed to them.

Pick a prize

Named in honour of the original host of the American game show *Let's Make a Deal*, the **Monty Hall problem** is a good example of maths at its most counter-intuitive. Problems such as this one have been around for some time. Parisian mathematician

Joseph Bertrand (1822–1900) posed a very similar problem called Bertrand's Box way back in 1889.

On *Let's Make a Deal*, the winner of the game show was offered the opportunity to choose one box out of three. One box contained the grand prize – the keys to a brand-new car. The other two boxes contained consolation prizes of much less worth.

The contestant picked a box but was forbidden from opening it. The host, Monty Hall, then revealed that one of the two remaining boxes was a consolation prize. He gave the contestant the option to swap their box for the remaining closed box. The question was:

Should the contestant swap boxes or not?

On the surface it seems that it should not matter. After all, the contestant is choosing between his or her own box and the other box, so it's a 50:50 call, right?

Wrong.

There are several ways of thinking about this problem and perhaps the most obvious way is to look at all possible outcomes of these events. Let's call the boxes A, B and C and say that A contains the prize.

Initial Choice	Stick or Swap	Outcome
A	Stick	Win
A	Swap	Lose
B	Stick	Lose
B	Swap	Win
C	Stick	Lose
C	Swap	Win

If we look at the outcome column, we can see that there are three ways to win and three ways to lose. So it was 50:50 after all!

Wrong.

If you look, one of the wins happens when you stick, and two of them happen when you swap. This means that the chance of winning if you swap is ⅔, and only ⅓ if you stick.

Another, less laborious way of thinking about it is this: when you pick initially, you have a 1 in 3 chance of choosing the correct box, which means there is a 2 in 3 chance that the prize box is one of the other two boxes. When the consolation box is revealed, it does not change the fact that there is a 2 in 3 chance that the remaining box is the prize.

A DIFFERENT LANGUAGE

We saw earlier how Alan Turing and other code breakers helped the Allies during the Second World War, but the discipline of **cryptography** precedes their efforts. People have been using codes and ciphers to communicate privately for thousands of years.

A **code** is, in effect, a language, and they work on the principle that A and B can communicate privately in a language that C does not know. The disadvantages here are that C can either find someone else who speaks the language, or they can find a dictionary or codebook and work out the information themselves.

Native Tongue

During the Second World War the USA used speakers of the native American dialect Navaho for secure radio communications because no Navaho dictionaries or non-American native speakers existed at the time.

Ciphers, on the other hand, involve altering the original message using a sort of algorithm; the art of the cryptologist is to break the cipher, often using mathematical means.

A new alphabet

The Roman army used Caesar's cipher to encrypt orders. This simple cipher worked by taking the letter in the original or **plaintext** and shifting each letter in it along the alphabet by a predetermined amount to create the **ciphertext**. For example, ATTACK would become BUUBDL if each letter is shifted along one place.

Although, admittedly, it is not a terribly cunning device, it is nevertheless effective if the orders were very short term. The problem here is that if you can work out the shift, you can translate the whole message easily.

More effective would be to rewrite the alphabet in a more random fashion, assigning each letter a new cipher letter without a constant shift. I could write A as Q, T as G, C as X and K as Z, so ATTACK would become QGGQXZ. This cipher is a bit more secure, but maybe vulnerable to attack if I set my computer the task of trying out different combinations of the alphabet. Let's see:

For A, I have 26 choices of letter (it might help to confuse things if sometimes the plaintext letter and the ciphertext are the same).

For B I have 25 choices, 24 for C, 23 for D, etc. This means there are $26 \times 25 \times 24 \times 23 \times \ldots \times 3 \times 2 \times 1$ possible alphabets for the computer to check. The shorthand for this is written 26! or 26 factorial.

$$26! \approx 403291461000000000000000000$$

This is a large number, but computers these days are so fast they can deal with a number of this size quite easily. Let's say my computer can check 100 alphabets per second to see if they make sense, so it would take my computer:

$$26!/100 \approx 4032914610000000000000000 \text{ seconds}$$
$$\approx 170488664000000000000000 \text{ years}$$

to check all the possible alphabets a message could be in. Clearly such a brute force approach would not work, but people have been able to break such ciphers for hundreds of years. How?

Language patterns

Arab mathematician Al-Kindi (801–73) developed something we now call **frequency analysis**. Al-Kindi discovered that every language uses its letters in unique proportions. For example, in English we use the letter e most often – 13% of the time, followed by t (9%), a (8%) and o (7.5%) and so on. Different languages have different percentages of letter distribution.

This means that, if you know an encrypted message is in English, you can count how many times each letter appears in the ciphertext and match it up with the real letters. For example, if the letter n appears 13% of the time it is most likely representing e. There are also idiosyncrasies in languages that help – such as q, which in English is nearly always followed by the letter u.

There are many other ciphers, each of which has become increasingly elaborate and difficult to break. However, in today's online world we have started transmitting far more of our private information, to the point where sending personal details such as bank account details, dates of birth and passwords is commonplace. So how do we protect this information?

Keys to the lock

In the nineteenth century the Dutch cryptographer Auguste Kerckhoff (1835–1903) summed up what makes an ideal cipher: even if someone knows the full workings of your cipher, it should still be impossible for them to decipher your message. This relies on having a good cipher and something called a **key**, which is the crucial information without which you are unable to break the cipher. In the example of Caesar's cipher, the key is how many letters along the alphabet you should shift.

Central to the workings of modern ciphers that control e-commerce, and also for people that like to keep their emails protected, are **one-way functions**, so-called because it's a calculation that is easy to do one way, but very difficult to do in reverse. British economist William Jevons (1835–82) recognized that it is fairly easy to multiply two prime numbers together; much harder is to work backwards from the result to find out which two prime numbers had been multiplied together, especially if the numbers are large.

For example, in a matter of moments you can multiply 23 and 19 using a calculator to get 437. However, to work out which prime numbers multiply to make 437 you have to go through the rather laborious process of checking whether each prime number goes into 437. In this case, I would have to perform a check for 2, 3, 5, 7, 11, 13, 17 before getting to 19. Checking seven times is not a big deal, but imagine checking prime numbers with thousands of digits. The largest primes known at the moment have over 12 *million* digits. This will take you a very long time, even using a computer.

In effect we now have two keys: the product of the two primes, which can be used to encrypt information, and the two prime numbers themselves, which can be used to decipher anything encrypted with the product of the prime numbers in the first place. This pair of keys is referred to as **public and private keys**. They allow information to be transmitted safely, even though people can get access to our encrypted data and they know what method we used to encrypt it.

Security Guard

A common analogy for this system, which is called **public key cryptography**, is normal 'snail' mail. You can make where you live common knowledge (the public key) and people can put messages through your door. But only *you* have the key to your house (the private key) to get in and read the message. The security in this example is your sturdy front door and excellent lock, which could be broken into but would take a prohibitively long time, just as it would take a computer a very long time to work out the original prime numbers that make up the private key.

So, when you buy something online, the website's computer lets your computer know its public key – a whopping great big number that is the sum of two big primes multiplied together. Your computer uses that number to encrypt all your details (which, if they weren't already numbers, have been converted to numbers) and then sends it to the website. The website's computer can then use its private key – the two prime numbers used in the first place – to decipher the message and remove the money from your account. Clever, eh?

GOING LARGE

When it comes to big numbers we have adopted the American system. Starting from 1 million, which originally derives from Italian, we then use Latin numbers as prefixes to indicate a number a thousand times larger. For example, 1 billion is 1,000 million, 'bi' being a prefix meaning two; 1 trillion is 1,000 billion, and 1 quadrillion is 1,000 trillion, and so forth.

The big and the small

The practical use for these numbers is limited to fields that deal with very large things (such as astronomy and cosmology) or very small things (such as chemistry and physics). For example, **Avogadro's constant** tells us how many atoms or molecules there are in a standard measure of a particular element, and that it is roughly equal to 600 sextillion. The mass of the earth is approximately equal to 6 septillion kilograms.

Mathematicians have another system with large and small numbers called **standard form** or **scientific notation**. This system exploits the fact that a number with n zeros after it is the same as the number multiplied by 10^n. Hence, Avogadro's constant is often written 6×10^{23}, which is much more convenient to use in calculations, and to enter into a calculator too. Standard form can also be convenient for very small numbers, because multiplying a number by a negative power of ten is the same as putting zeros and a decimal point in front of the number. For example, an electron's mass is approximately 9×10^{-31} kg, e.g. a 9 preceded by 31 zeros with a decimal point between the first two: 0.0000000000000000000000000000009.

Where Google Got Its Name

When American mathematician Edward Kasner (1878–1955) wanted to create new words for large numbers he asked his young nephew for help. The boy suggested the word 'googol', which would have a value of 1 followed by 100 zeros:

10000000000000000000000000000000000000000000000000000000000000000000
0000000000000000000000000000000000000

The pair also quickly invented the 'Googolplex', which is 1 followed by a googol zeros, or 10^{googol}. Kasner's motivation here was to demonstrate that you could have these incredibly large numbers that still were not infinite.

The founders of Google used a modified version of the word googol to imply that their search engine could sift through a very large number of websites quickly.

PARTY TIME

In 1939 Austrian engineering scientist Richard von Mises (1883–1953) posed the **Birthday problem**: how many people need to be in a room for there to be a 50% chance of two of them sharing a birthday? The answer to this puzzle is surprising.

Most people's first response to the conundrum goes something along the lines of: there would need to be 366 people in the room to guarantee 2 of them sharing a birthday, so 183 people (half of 366) in the room would give a 50% chance.

The correct answer, however, is only 23 people. Although this seems very unlikely, it is in fact true. Why?

Well, if there are 2 people in the room, Alan and Blaise, there is only one way they can both share a birthday. If a third person, Carl, joins them, then there are three possible matches:

AB BC

AC

If a fourth person, Delia, joins them, the possible matches are:

AB BC CD

AC BD

AD

Now there are six chances for two people in the room to share a birthday. Each time a new person joins the room another layer is added to the triangle:

AB BC CD DE

AC BD CE

AD BE

AE

With five people in the room there are ten combinations that could result in a shared birthday. This triangular pattern stops you from having to add further layers because the **triangular numbers** that the pattern produces are well known to mathematicians.

Even more triangles!

Triangular numbers are formed by adding together consecutive whole numbers. Hence, the first triangular number is 1, the second is 1 + 2 = 3, the third is 1 + 2 + 3 = 6 etc.

*	*	*	*	*
	**	**	**	**
		***	***	***
			****	****
				*****
1	3	6	10	15

The formula for the nth triangular number is $\frac{1}{2} \times n \times (n+1)$, but I can also find the triangular numbers using Pascal's triangle (see page 82), by looking down the third diagonal:

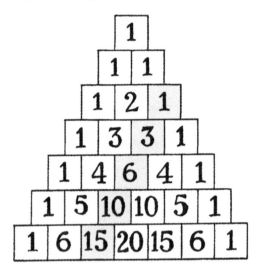

The triangular numbers also have a role to play in the carol 'The Twelve Days of Christmas'. In this song, each day your 'true love' gives to you an ever-increasing quantity of gifts:

Day	Gift	Number of Gifts	Total
1	1 partridge in a pear tree	1	1
2	2 turtle doves	3	4
	1 partridge in a pear tree		
3	3 French hens	6	10
	2 turtle doves		
	1 partridge in a pear tree		
4	4 calling birds	10	20
	3 French hens		
	2 turtle doves		
	1 partridge in a pear tree		

As you can see, the number of gifts awarded each day corresponds to the triangular numbers. But what about the running total of gifts? These numbers are the next diagonal of Pascal's triangle and are known as the **tetrahedral numbers**, because each time you add another day you are adding another layer to the pyramid:

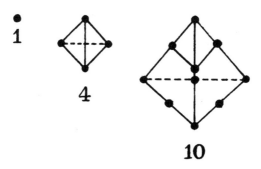

The formula for these numbers is $n \times (n+1) \times (n+2) \div 6$. To find the total number of gifts your generous true love has given to you, set the following:

n = 12

$12 \times 13 \times 14 \div 6 = 364$

That's a pretty decent haul!

Return to the party

Back to the Birthday problem: in order to find out the number of possible combinations of birthdays shared between 2 people in a group of 23 people you need to know the 22nd triangular number:

$\frac{1}{2} \times 22 \times 23 = 253$ combinations

With this many pairings of 23 people, it now seems more reasonable that there is a 50% chance of two of them sharing a birthday.

To show the exact probability here, it is actually easier to find the opposite – the chance that *no* two people share a birthday – and exploit the fact that in probability the chance of something happening and the chance of something not happening have a sum of 100%.

Alan's birthday can fall on any of the 365 days of the year, leaving 364 alternative days on which Blaise's birthday could fall. In turn, there are 363 days on which Carl's birthday could fall in order for it not to be shared with either Alan or Blaise. By the time the 23rd guest, Walter, steps into the room there are 343 possible days on which *his* birthday could fall without it being shared by anyone else in the room. If you write each one of these numbers as a probability out of 365, and then multiply them together, the total probability generated is:

$$\frac{(365 \times 364 \times 363 \times 362 \times \ldots \times 345 \times 344 \times 343)}{365^{23}} = 49.3\%$$

which means that the probability of 2 people sharing a birthday when there are 23 people in the room is 50.7%.

THE ALIENS HAVE LANDED

In 1960 American astronomer Frank Drake (1928–) was the first person to use radio telescopes to search for signals, messages or other evidence of intelligent life in the universe. This spawned what is now known as **SETI**– the Search for Extra-Terrestrial Intelligence.

Drake developed an equation to calculate the number of civilizations in the Milky Way that we should be able to communicate with:

$$\text{number of civilizations} = R^* \times f_p \times n_e \times f_l \times f_i \times f_c \times L$$

R^* is the number of new stars made in the galaxy each year; f_p is the fraction of new stars that will have planets; n_e is the number of potentially life-supporting planets; f_l is the fraction of the life-supporting planets that are known to have life on them; f_i is the fraction of planets that have intelligent life; and f_c is the fraction of planets that emit some kind of evidence of their civilization, such as radio waves. L represents how long such evidence emits for.

At the time of writing, many of these factors are pure conjecture, and scientists have come up with widely varying answers. Why not try your own!

A Foreign Language

It has been suggested that, if we do make contact with an alien civilization, numbers may be one of the first ways in which we communicate. In 1974 the Arecibo radio telescope in Puerto Rico beamed a radio message in the direction of a galaxy 25,000 light years away. Much of the information contained was numerical: the numbers from 1 to 10, the atomic numbers of the elements that make up DNA, the height of a man and the population of the earth.

The Future of Mathematics

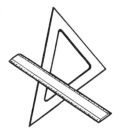

We're not done yet. Of course, the use of new computer methods to solve numerical problems that were previously deemed impossible to solve has boosted enormously developments in technology, science, medicine and engineering. However, there still remain thousands of unsolved problems in mathematics and science that will keep the experts busy for some time to come...

THE MARCH FORWARD

In 1900, at the International Congress of Mathematicians, German mathematician David Hilbert (1862–1943) posed twenty-three mathematical problems that he felt were key to the development of the subject. Since the congress ten of Hilbert's problems have been solved, seven have been solved to some extent or have been shown not to have a solution, three were too vague to be solved and three remain unsolved.

Posing these problems had exactly the effect that Hilbert wanted – the competition spurred mathematicians to strive to tackle them and in the process forge into new areas of research. In 2000, in much the same vein as Hilbert, the Clay Mathematics Institute issued another seven problems, now known as the Millennium problems.

So far, only one of the problems has been solved – the Poincaré conjecture, which relates to the topology of spheres. It was solved by an extraordinary Russian mathematician called Grigori Perelmann, who has declined not only a Fields Medal (the highest accolade in mathematics) but also the $1 million prize from the Clay Institute.

One of the problems posed in both Hilbert's problems and the Millennium problems is the Riemann hypothesis, a problem that many mathematicians feel is the most important in mathematics. It concerns the distribution of prime numbers. The Goldbach conjecture (see page 132) tells us approximately where the prime numbers should be; the Riemann hypothesis would help us to know how far away from the expected place the prime should actually be.

WHAT NEXT?

The future of mathematics depends very much on mathematicians who are, as I write, children, or as yet unborn. In order to cultivate the best possible mathematicians and scientists to help solve the world's problems we need people with excellent mathematical training, which is quite an educational investment. In our current educational system, every schoolchild is taught numbers and arithmetic through to algebra and geometry so that by the onset of adulthood they have the tools necessary to enter a technical career path, should they so choose.

The majority of people, however, do not enter a technical career and therefore do not necessarily need mathematics taught beyond primary school. Most people use calculators or, more frequently, mobile telephones with built-in calculators, to do the mundane arithmetic that is all the maths needed in everyday life.

So, should we continue to make mathematics a compulsory subject until the age of sixteen? There are clearly those who enjoy maths and those who do not. Perhaps we could just teach basic arithmetic and everyday maths to younger children and save the harder, more interesting stuff as an optional course for older children who show a particular inclination and aptitude towards the subject? Well, if it worked for the ancient Greeks...

The fundamental theories of how the universe works – as discovered by, among others, Newton, Einstein, Feynman and Hawking – have been made through the creation of a mathematical model and the pursuit of the mathematical conclusions that ensue. These models are then tested by experiments in the real world to check the accuracy of the model.

As time goes by, it seems that in order to generate the best possible rate of advancement in science, we need mathematicians who understand the latest developments in that field, and scientists who understand the latest developments in mathematics too.

As Galileo said:
'Mathematics is the language with which God has written the universe.'

Bibliography

50 Mathematical Ideas You Really Need to Know
by Tony Crilly (Quercus)
A History of Mathematics
by Carl Boyer and Uta Merzbach (John Wiley & Sons)
Descartes: Key Philosophical Writings
by Rene Descartes trans. Elizabeth Haldane and
G. Ross (Wordsworth Classics)
Elements of Geometry by Euclid trans. Richard Fitzpatrick
Fermat's Last Theorem by Simon Singh (Fourth Estate)
How Mathematics Happened: The First 50,000 Years
by Peter Rudman (Prometheus)
Mathematics: From the Birth of Numbers
by Jan Gullberg (WW Norton & Co.)
Number: From Ancient Civilisations to the Computer
by John McLeish (Flamingo)
The Code Book by Simon Singh (Fourth Estate)
The History of Mathematics: A Very Short Introduction
by Jacqueline Stedall (OUP)
The Psychology of Learning Mathematics
by Richard Skemp (Pelican)

Index